Lecture Notes in Computer Science   13797

The series Lecture Notes in Computer Science (LNCS), including its subseries Lecture Notes in Artificial Intelligence (LNAI) and Lecture Notes in Bioinformatics (LNBI), has established itself as a medium for the publication of new developments in computer science and information technology research, teaching, and education.

LNCS enjoys close cooperation with the computer science R & D community, the series counts many renowned academics among its volume editors and paper authors, and collaborates with prestigious societies. Its mission is to serve this international community by providing an invaluable service, mainly focused on the publication of conference and workshop proceedings and postproceedings. LNCS commenced publication in 1973.

Moi Hoon Yap · Connah Kendrick · Bill Cassidy
Editors

Diabetic Foot Ulcers Grand Challenge

Third Challenge, DFUC 2022
Held in Conjunction with MICCAI 2022
Singapore, September 22, 2022
Proceedings

Editors
Moi Hoon Yap 🆔
Manchester Metropolitan University
Manchester, UK

Connah Kendrick 🆔
Manchester Metropolitan University
Manchester, UK

Bill Cassidy 🆔
Manchester Metropolitan University
Manchester, UK

ISSN 0302-9743   ISSN 1611-3349 (electronic)
Lecture Notes in Computer Science
ISBN 978-3-031-26353-8   ISBN 978-3-031-26354-5 (eBook)
https://doi.org/10.1007/978-3-031-26354-5

This Springer imprint is published by the registered company Springer Nature Switzerland AG
The registered company address is: Gewerbestrasse 11, 6330 Cham, Switzerland

Preface

This is the second Diabetic Foot Ulcer (DFU) proceedings, this one focusing on the DFU Challenge 2022 (DFUC 2022), organized in conjunction with the 25th International Conference on Medical Image Computing and Computer Assisted Intervention (MICCAI 2022). Following two challenges (DFUC 2020 and DFUC 2021) conducted as virtual events due to the COVID-19 pandemic, this challenge event was conducted in person on September 22, 2022 in Singapore. We received over 1000 submissions for the validation stage and 52 submissions for the testing stage.

The DFU challenges aim to motivate the healthcare domain to share datasets, participate in ground truth annotation, and enable data-innovation in computer algorithm development. In the longer term, it will lead to improved patient care and reduce the strain on overburdened healthcare systems. With joint efforts from leading scientists from the UK, US, India and New Zealand, these inaugural challenges were well-received, with the first DFU challenge successfully conducted in 2020 for the task of DFU detection and the second DFU challenge successfully conducted in 2021 for the task of DFU classification. The task of DFUC 2022 was ulcer segmentation, for the purpose of supporting research towards more advanced methods of ulcer region quantification.

This proceedings includes research papers investigating the effect of duplicate images in DFUC 2021 (as a continuation from the DFUC 2021 proceedings), gathers methodological papers of segmentation methods evaluated at DFUC 2022 and post challenge, and includes a DFUC 2022 summary paper from the organizers. Apart from the continuation paper of DFUC 2021 and a summary paper, all papers were reviewed by 2–3 reviewers and assigned to a meta-reviewer. For DFUC 2022 papers a single-blind review process was used, accepting only the top-5 entries due to the proceedings call for papers deadline. In total we received 21 paper submissions with only eight papers accepted and two withdrawn with an acceptance rate of 45%. The organizers were not listed as authors for the challenge and post challenge papers.

As a concluding note, the organizers of the challenge continue to support the research community by providing live leaderboards to test the performance of their algorithms. To date, there are 204 submissions on the DFUC 2022 live leaderboard. Researchers who are interested in the challenge can request the datasets and evaluate on our grand challenge websites (DFU detection task, DFU classification task and DFU segmentation task). The organizers will conduct DFUC 2023 in conjunction with MICCAI 2023, with the task focusing on DFU instance segmentation.

December 2022

Moi Hoon Yap
Connah Kendrick
Bill Cassidy

Organization

General Chairs

Moi Hoon Yap   Manchester Metropolitan University, UK
Connah Kendrick   Manchester Metropolitan University, UK
Neil Reeves   Manchester Metropolitan University, UK

Organizing Committee

Moi Hoon Yap   Manchester Metropolitan University, UK
Neil Reeves   Manchester Metropolitan University, UK
Andrew Boulton   University of Manchester and Manchester
     Infirmary, UK
Satyan Rajbhandari   Lancashire Teaching Hospitals, UK
David Armstrong   University of Southern California, USA
Arun G. Maiya   Manipal College of Health Professions, India
Bijan Najafi   Baylor College of Medicine, USA
Bill Cassidy   Manchester Metropolitan University, UK
Justina Wu   Waikato District Health Board, New Zealand

Clinical Chairs

Joseph M. Pappachan   Lancashire Teaching Hospitals, UK
Claire O'Shea   Waikato District Health Board, New Zealand

Technical Chairs

Connah Kendrick   Manchester Metropolitan University, UK
Bill Cassidy   Manchester Metropolitan University, UK

Program Committee Chairs

Connah Kendrick   Manchester Metropolitan University, UK
Bill Cassidy   Manchester Metropolitan University, UK

Program Committee

Christoph Friedrich	University of Applied Sciences and Arts Dortmund, Germany
Raphael Brüngel	University of Applied Sciences and Arts Dortmund, Germany
Nora Al-Garaawi	University of Kufa, Najaf, Iraq
Adrian Galdran	Bournemouth University, UK
Ting-Yu Liao	National Tsing Hua University, Taiwan
Joanna Jaworek-Korjakowska	AGH University of Science and Technology, Poland
Andrzej Brodzicki	AGH University of Science and Technology, Poland
Yung-Han Chen	National Yang Ming Chiao Tung, Taiwan
David Hresko	Technical University of Kosice, Slovakia
Abdul Qayyum	École Nationale d'Ingénieurs de Brest, France

Sponsors

Contents

Summary Paper

Quantifying the Effect of Image Similarity on Diabetic Foot Ulcer Classification

Imran Chowdhury Dipto[1], Bill Cassidy[1], Connah Kendrick[1],
Neil D. Reeves[3], Joseph M. Pappachan[2], Vishnu Chandrabalan[2],
and Moi Hoon Yap[1(✉)]

[1] Centre for Advanced Computational Science, Department of Computing
and Mathematics, Manchester Metropolitan University, Manchester M1 5GD, UK
M.Yap@mmu.ac.uk
[2] Lancashire Teaching Hospitals NHS Trust, Preston PR2 9HT, UK
[3] Musculoskeletal Science and Sports Medicine Research Centre, Manchester
Metropolitan University, Manchester M1 5GD, UK

Abstract. This research conducts an investigation on the effect of visually similar images within a publicly available diabetic foot ulcer dataset
when training deep learning classification networks. The presence of
binary-identical duplicate images in datasets used to train deep learning algorithms is a well known issue that can introduce unwanted bias
which can degrade network performance. However, the effect of visually
similar non-identical images is an under-researched topic, and has so far
not been investigated in any diabetic foot ulcer studies. We use an open-
source fuzzy algorithm to identify groups of increasingly similar images in
the Diabetic Foot Ulcers Challenge 2021 (DFUC2021) training dataset.
Based on each similarity threshold, we create new training sets that we
use to train a range of deep learning multi-class classifiers. We then evaluate the performance of the best performing model on the DFUC2021 test
set. Our findings show that the model trained on the training set with the
80% similarity threshold images removed achieved the best performance
using the InceptionResNetV2 network. This model showed improvements
in F1-score, precision, and recall of 0.023, 0.029, and 0.013, respectively.
These results indicate that highly similar images can contribute towards
the presence of performance degrading bias within the Diabetic Foot
Ulcers Challenge 2021 dataset, and that the removal of images that
are 80% similar from the training set can help to boost classification
performance.

1 Introduction

Since the publication of the DFUC2021 Proceedings, there has been no sub-
stantial progress made on the DFU multi-class classification task. This paper
studies one of possible cause, i.e., the effect of visually similar non-identical
images within DFUC2021 dataset. Image duplication (the presence of binary

---

I. C. Dipto and B. Cassidy—Equal contribution.

identical images) is generally acknowledged as a factor in reducing model performance when training deep learning models due to the performance degrading bias that over-represented features may introduce into the trained model. However, the effect of visually similar non-identical images which may result in an over-representation of certain features present in deep learning datasets is an under-researched topic. An overabundance of certain features in a dataset may cause undesirable performance degrading bias in any models trained using them. In this paper we conduct an analysis of the effect of images that are visually similar but not binary identical on a publicly available diabetic foot ulcer dataset using an open-source fuzzy matching algorithm. We train a large range of multi-class deep learning classification models on the Diabetic Foot Ulcer Challenge 2021 dataset (DFUC2021) [1], and for the first time, quantify the effect of image similarity on network accuracy.

We found no studies that observed and quantified the effect of feature over-representation or the effect of image similarity in DFU research. The effect of binary duplicate images has been observed in other domains [2] but the topic remains an under-researched problem generally. A common theme with many previous studies is limited dataset size. A small dataset may hinder the ability of models to generalise to a wider range of examples in real-world settings. Conversely, a large dataset might introduce performance degrading feature-bias if the data has been collected from a small number of subjects. To address this, we conduct experiments to analyse the effect of image similarity on the DFUC2021 dataset.

2   Related Work

Previous research on DFU has involved localisation [3–8], binary [9] and multi-class classification [10,11], and segmentation [12,13]. Goyal et al. [14] proposed a deep learning architecture for DFU classification. Their work was notable for achieving high scores in sensitivity and accuracy using a small DFU dataset (<2000 images for training and testing).

More recently, Al-Garaawi et al. [15] conducted a series of binary classification experiments using DFU patches. This work used mapped local binary pattern coded images with RGB images as inputs to the CNN to increase binary classifier performance. However, a limitation of this work is the use of small datasets.

In last year's Diabetic Foot Ulcers Grand Challenge, Yap et al. [1] conducted multi-class classification experiments using the DFUC2021 dataset. This work highlighted the challenging nature of multi-class classification in this domain due to intra-class similarities.

3   Methodology

To observe the effect of similar images in the DFUC2021 dataset on multi-class classification, we devised a strategy of gradually removing successive groups of similar images from the training set. Each group of similar images were identified using the dupeGuru [16] Windows application. This open-source application implements a fuzzy search algorithm capable of identifying visually similar images. Results can be filtered by percentage similarity within the application. Using this feature, we were able to identify groups of similar images within the DFUC2021 training set. For each similarity threshold, we train a set of multiclass classifiers capable of classifying the following five classes: (1) control, (2) infection, (3) ischemia, (4) infection and ischemia, and (5) none. Figure 1 shows an overview of the entire process used to create the new training sets used in our experiments.

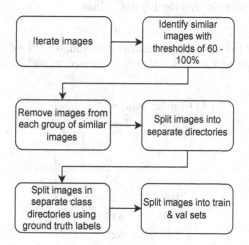

Fig. 1. Overview of the process used in the identification and removal of similar images at each similarity threshold.

3.1   Dataset Description

For our experiments, we use the publicly available DFUC2021 dataset, introduced by Yap et al. [1]. This dataset is the largest publicly available DFU dataset with wound pathology class labels. The dataset comprises a total of 15,683 images, sized at 224 × 224 pixels, with 5955 images for the training set (2555 infection only, 227 ischaemia only, 621 both infection and ischaemia, and 2552 without ischaemia and infection), 3994 unlabeled images, and 5734 images for the testing set. All wounds are cropped from larger images so that only the wound is present in each image. The DFUC2021 dataset is highly heterogeneous due to the nature of the variety of capture devices and variable settings used during photographic acquisition. The ground truth labels were provided by expert clinicians at Lancashire Teaching Hospitals, UK. This dataset was obtained with ethical approval from the UK National Health Service Research Ethics Committee (reference number: 15/NW/0539).

3.2   Fuzzy Algorithm

The fuzzy algorithm used by the dupeGuru application reads each image in RGB bitmap mode which is then split into blocks. Next, the analysis phase uses a 15×15 pixel grid to average the colour of each grid tile, the results of which are stored in a cache database. Each grid tile, representing an average colour, is then compared to its corresponding grid on the other image being compared to, and a sum of the difference between R, G and B on each side is computed. The RGB sums are then added together to obtain a final result. If the score is smaller or equal to the user-specified threshold, then a match is found. If a threshold of 100 is set by the user then the algorithm adds an extra constraint indicating that images should contain identical binary data.

Table 1. Summary of the number of similar images found by the dupeGuru fuzzy algorithm in the train, test, and train & test sets combined at each user-defined similarity threshold.

Threshold (%)	Similar images		
	Train set	Test set	Train & Test set
100	0	0	0
95	1	0	3
90	19	23	48
85	106	125	268
80	317	345	719
75	590	621	1278
70	1013	979	2082
65	1509	1367	2976
60	2066	1683	3906

3.3   Identification of Similar Images

To identify the similar images in the DFUC2021 dataset we have used the hardness level (similarity threshold) in the dupeGuru application with the values of 60%, 65%, 70%, 75%, 80%, 85%, 90%, 95%, and 100%. A similarity threshold of 80%, for example, indicates that the application will find images that have 80% similarity. We ran the fuzzy algorithm on the training, test, and the training and test sets combined to find similar images that exist exclusively within the training set, exclusively within the test set, and within both train and test sets combined. Table 1 shows a summary of the number of similar images detected by the dupeGuru fuzzy algorithm on the different thresholds in the training set, the test set, and the training and test sets combined.

3.4   Removal of Similar Images as Determined by Similarity Thresholds

Images of each of the classes present in the dataset were separated into different directories - one directory per class. On each of these directories the dupeGuru fuzzy algorithm was run using similarity thresholds of 60% to 100%. Next, the filenames for the images in each similarity threshold were saved into CSV files containing Group ID and Image filenames. Group ID refers to the grouping of similar images returned by the fuzzy algorithm, where each group of similar images is assigned a unique sequential identifier. The CSV files output by dupe-Guru were then merged with the file containing the ground truth labels of the training set based on the image filenames. For each group of similar images, all but the first image in each group was removed. This meant that a single example from each similarity group was kept for inclusion in each similarity threshold training set.

Table 2. Summary of the number of similar images removed at each similarity threshold and the remaining images that are used for the new training sets.

Threshold (%)	Similar images removed	Remaining images
95	1	9948
90	19	9930
85	106	9843
80	317	9632
75	590	9359
70	1013	8936
65	1508	8441
60	2068	7881

By comparing the filenames with the ground truth labels, the images in the curated datasets were copied into new directories which formed the new training sets. To check the validity of the results from this process, a Python routine was created which compared the CSV files against the ground truth labels to ensure that the correct images had been copied and that non of the additional images from each similarity group were present in the new training sets. An additional manual spot-check was completed on a random sample of images in the new training sets to ensure that the images had been correctly separated. Table 2 shows a summary of the number of images removed at each similarity threshold together with the total remaining images used to form the new training sets.

4   Image Similarity Analysis

In this section we analyse a selection of images from each of the similarity thresholds returned by the dupeGuru fuzzy algorithm prior to training the multi-class classification models. Note that not all similarity searches returned results.

4.1   Train Set Image Similarity

Figure 2(a & b) shows two images from the training set in the 75% similarity threshold. These images are of the same wound at different levels of magnification, with the example shown in (a) being at a higher level of magnification. Figure 2(c & d) shows two training set images in the 65% similarity threshold. These two images represent two distinctly different DFU wounds with noticeably different features. Figure 2(e & f) shows a further two training set images which were identified in the 60% similarity threshold. As with the previous examples, these images represent two different wounds.

4.2   Test Set Image Similarity

Figure 3(a & b) shows two similar images from the 80% threshold. These images are of the same wound, with the second image being a natural augmentation case with a slightly different zoom level. The main visual differences can be observed on the bottom section of the image where the dark spots in image (a) are not present on image (b). Figure 3(c & d) shows two similar images from the test set with 65% similarity. As per the previous test examples (Fig. 3(a & b)), the second image represents a case of natural augmentation where the wound has a noticeably increased zoom level.

4.3   Train and Test Set Image Similarity

Figure 4(a & b) shows two distinctly different wound images identified in the 75% similarity threshold. Image (a) is from the training set, image (b) is from the test set. Figure 4(c & d) and (e & f) show further examples of visually similar images found across training and test sets at 65% and 60% similarity thresholds respectively. Note that we do not discuss the class of the test set examples as these are part of a live public challenge for DFUC2021 which is still open to submissions (https://dfu-challenge.github.io/dfuc2021.html).

4.4   Inter-class Image Similarity

Our experiments using the dupeGuru fuzzy algorithm did not return any inter-class similarity results for the following groups of classes: (1) 'both' vs 'none', and (2) 'infection' vs 'ischemia'. For the 'infection' vs 'none' similarity searches, similar images were found for the 70% (45 images), 75% (13 images), and 80% (4 images) similarity brackets. For the 'ischemia' vs 'none' similarity searches,

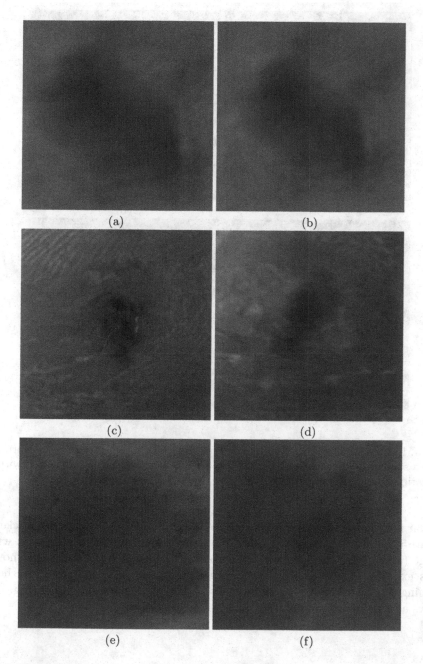

Fig. 2. Illustration of two training set images identified by the dupeGuru fuzzy algorithm with a similarity threshold of 75% (a & b), two training set images from the 'none' class found in the 65% similarity threshold (c & d), and two training set images found in the 60% similarity threshold - (e) is from the 'none' class, (f) is from the 'unlabelled' class.

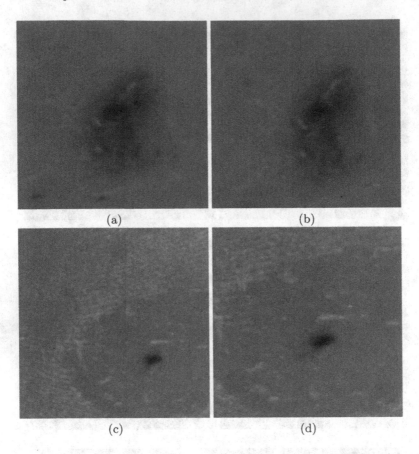

(a)   (b)

(c)   (d)

Fig. 3. Illustration of two test set images identified by the dupeGuru fuzzy algorithm with a similarity threshold of 80% (a & b), and two test set images identified with a similarity threshold of 65% (c & d).

similar images were found for the 70% (5 images) and 75% (3 images) brackets. Figure 5 shows two images from the 75% similarity threshold training set, with image (a) showing an 'infection' class example, and image (b) showing a 'none' class example. Figure 6 shows two images from the 70% similarity threshold training set, where image (a) shows an 'ischemia' class example, while image (b) shows a 'none' class example.

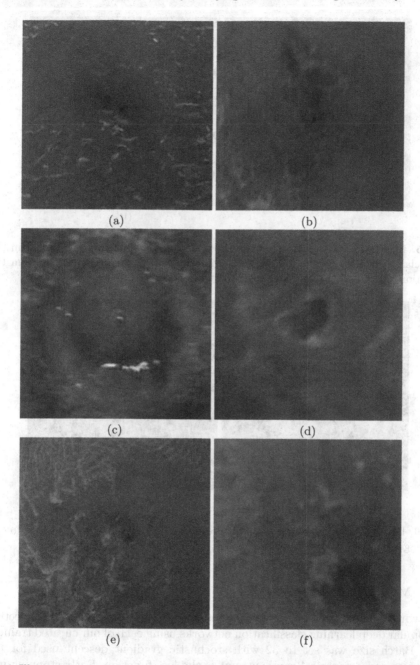

Fig. 4. Illustration of similar images located across both training and test sets identified in the 60% to 75% similarity thresholds. Image (a) shows a training set image and image (b) shows a test set image, both identified in the 75% similarity threshold. Image (c) shows a training set image and image (d) shows a test set image, both identified in the 65% similarity threshold. Image (e) shows a training set image and image (f) shows a test set image, both identified in the 60% similarity threshold.

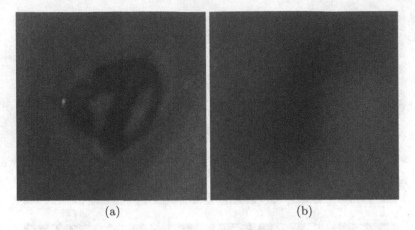

(a)   (b)

Fig. 5. Illustration of two images from the training set which were identified in the inter-class similarity results for the 75% similarity threshold: (a) an image from the 'infection' class, and (b) an image from the 'none' class

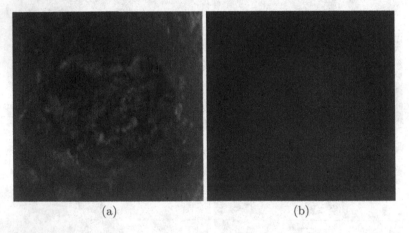

(a)   (b)

Fig. 6. Illustration of inter-class similarity results in the 70% similarity threshold: (a) an image from the 'ischemia' class, and (b) an image from the 'none' class.

4.5   Model Training

Following the creation of each training set, as per Table 2, we trained a selection of popular deep learning classification networks using each of our curated training sets. Batch size was set to 32 with stochastic gradient descent used for the optimiser and categorical cross-entropy as the loss function. Early stopping was implemented monitoring validation accuracy with a patience of 10. The hardware configuration used for our experiments was as follows: Intel Core i7-10750H CPU @2.60 GHz, 64 GB RAM, NVIDIA GeForce RTX 2070 Super with Max-Q Design 8 GB. The software configuration used was as follows: Ubuntu 20.04 LTS, Python 3.8.13, and Tensorflow 2.4.2.

5   Results and Discussion

This section details the results of training the multi-class classification networks on each of the training sets as detailed in Table 2. We report the validation results from the full training set and the training sets using the following similarity thresholds: (1) 60%, (2) 65%, (3) 70%, (4) 75%, (5) 80%, (6) 85%, (7) 90%, and (8) 95%. Finally, we report the test results for the model with the highest validation accuracy.

5.1   Baseline Results

Table 3 shows the validation accuracy results for the models trained using the full training set, with no similar images removed. The InceptionResNetV2 model shows a clear lead in validation accuracy with 0.801, showing an increase of 0.038 over the next best performing model, which was ResNet50 with a validation accuracy of 0.763.

Table 3. Best epoch and validation accuracy of the models trained on the full DFUC2021 dataset.

Model	Best epoch	Validation accuracy
DenseNet201	36	0.731
EfficientNetB0	25	0.656
EfficientNetB1	11	0.587
EfficientNetB3	38	0.649
InceptionResNetV2	42	**0.801**
InceptionV3	16	0.704
ResNet50	47	**0.752**
ResNet50V2	43	0.738
ResNet101	30	0.719
ResNet101V2	54	0.735
ResNet152	19	0.703
ResNet152V2	76	**0.763**
VGG16	52	0.687
VGG19	75	0.716
Xception	18	0.735

5.2   Results on the Curated Datasets

The validation results for each of our curated training sets are shown in Table 4, Table 5, Table 6, Table 7, Table 8, Table 9, Table 10, and Table 11. The best performing model for validation accuracy is InceptionResNetV2 on the 80% similarity threshold (0.885), as shown in Table 8. This represents an increase of 0.030

over the next best performing model across all the best performing models in all similarity thresholds, which was InceptionResNetV2 in the 75% similarity threshold with 0.855 accuracy as shown in Table 7.

Table 4. Best Epoch and validation accuracy of the models trained on the 60% similarity threshold dataset. BE - best epoch, Va - validation accuracy.

Model	Best epoch	Validation accuracy
DenseNet121	2	0.399
DenseNet169	2	0.419
DenseNet201	3	0.403
EfficientNetB0	5	0.445
EfficientNetB1	4	0.424
EfficientNetB3	7	0.443
InceptionResNetV2	3	**0.470**
InceptionV3	3	**0.461**
ResNet50	3	0.431
ResNet50V2	8	0.416
ResNet101	9	0.410
ResNet101V2	1	0.421
ResNet152	2	**0.460**
ResNet152V2	2	0.432
VGG16	10	0.435
VGG19	14	0.442
Xception	5	0.416

The lowest performing models across all similarity thresholds for validation accuracy were ResNet152 (0.460), InceptionV3 (0.461), and InceptionResNetV2 (0.470), all present in the 60% similarty threshold (see Table 4). This indicates that the 60% similarity threshold removed too many useful examples that the models were able to learn from - 2066 images compared to 317 images for the best performing network (InceptionResNetV2 at 80% similarity threshold) in validation accuracy. All models trained in the 60% similarity threshold also show low convergence for best epoch when compared to all other similarity thresholds, further highlighting a lack of learnable features present in this heavily curated training set.

Given that the InceptionResNetV2 model trained on the 80% similarity threshold training set performed best in validation accuracy, we used this model to obtain test results on the DFUC2021 testing set. The results for this experiment are presented in Table 12. The InceptionResNetV2 model trained on the 80% similarity threshold training set shows clear performance improvements for

Table 5. Best Epoch and validation accuracy of the models trained on the 65% similarity threshold dataset.

Model	Best epoch	Validation accuracy
DenseNet121	19	0.673
DenseNet169	33	0.731
DenseNet201	21	0.715
EfficientNetB0	44	0.670
EfficientNetB1	53	0.664
EfficientNetB3	28	0.662
InceptionResNetV2	54	**0.823**
InceptionV3	37	**0.756**
ResNet50	54	**0.754**
ResNet50V2	35	0.742
ResNet101	32	0.729
ResNet101V2	17	0.671
ResNet152	26	0.713
ResNet152V2	55	0.744
VGG16	21	0.651
VGG19	42	0.662
Xception	26	0.733

Table 6. Best Epoch and validation accuracy of the models trained on the 70% similarity threshold dataset.

Model	Best epoch	Validation accuracy
DenseNet121	60	**0.765**
DenseNet169	34	0.707
DenseNet201	15	0.686
EfficientNetB0	30	0.693
EfficientNetB1	27	0.660
EfficientNetB3	7	0.581
InceptionResNetV2	29	**0.759**
InceptionV3	26	0.715
ResNet50	7	0.650
ResNet50V2	18	0.684
ResNet101	27	0.719
ResNet101V2	28	0.684
ResNet152	27	0.707
ResNet152V2	14	0.660
VGG16	28	0.669
VGG19	34	0.686
Xception	48	**0.791**

Table 7. Best Epoch and validation accuracy of the models trained on the 75% similarity threshold dataset.

Model	Best epoch	Validation accuracy
DenseNet121	27	0.734
DenseNet169	12	0.692
DenseNet201	16	0.684
EfficientNetB0	56	0.704
EfficientNetB1	17	0.634
EfficientNetB3	71	0.73
InceptionResNetV2	93	**0.855**
InceptionV3	18	0.717
ResNet50	24	0.706
ResNet50V2	45	**0.747**
ResNet101	32	**0.74**
ResNet101V2	19	0.675
ResNet152	28	0.706
ResNet152V2	61	0.743
VGG16	50	0.70
VGG19	58	0.712
Xception	17	0.734

Table 8. Best epoch and validation accuracy of the models trained on the 80% similarity threshold dataset.

Model	Best epoch	Validation accuracy
DenseNet121	8	0.662
DenseNet169	44	0.759
DenseNet201	20	0.704
EfficientNetB0	60	0.708
EfficientNetB1	15	0.619
EfficientNetB3	34	0.681
InceptionResNetV2	99	**0.885**
InceptionV3	43	**0.769**
ResNet50	43	0.765
ResNet50V2	32	0.722
ResNet101	46	0.747
ResNet101V2	86	**0.789**
ResNet152	45	0.733
ResNet152V2	38	0.736
VGG16	25	0.674
VGG19	22	0.672
Xception	30	0.765

Table 9. Best epoch and validation accuracy of the models trained on the 85% similarity threshold dataset.

Model	Best epoch	Validation accuracy
DenseNet121	15	0.688
DenseNet169	22	0.709
DenseNet201	27	0.742
EfficientNetB0	47	0.700
EfficientNetB1	14	0.589
EfficientNetB3	16	0.605
InceptionResNetV2	52	**0.805**
InceptionV3	28	0.707
ResNet50	53	0.748
ResNet50V2	64	**0.767**
ResNet101	35	0.717
ResNet101V2	50	0.735
ResNet152	61	**0.755**
ResNet152V2	22	0.683
VGG16	61	0.683
VGG19	44	0.682
Xception	13	0.715

Table 10. Best epoch and validation accuracy of the models trained on the 90% similarity threshold dataset.

Model	Best epoch	Validation accuracy
DenseNet121	20	0.693
DenseNet169	36	0.740
DenseNet201	14	0.700
EfficientNetB0	20	0.653
EfficientNetB1	40	0.658
EfficientNetB3	40	0.659
InceptionResNetV2	41	**0.785**
InceptionV3	57	**0.768**
ResNet50	35	0.736
ResNet50V2	23	0.699
ResNet101	46	**0.743**
ResNet101V2	10	0.665
ResNet152	28	0.715
ResNet152V2	10	0.649
VGG16	61	0.698
VGG19	61	0.689
Xception	6	0.669

Table 11. Best epoch and validation accuracy of the models trained on the 95% similarity threshold dataset.

Model	Best epoch	Validation accuracy
DenseNet121	78	**0.817**
DenseNet169	34	**0.740**
DenseNet201	24	0.720
EfficientNetB0	34	0.677
EfficientNetB1	31	0.664
EfficientNetB3	12	0.595
InceptionResNetV2	21	0.734
InceptionV3	28	0.726
ResNet50	46	0.737
ResNet50V2	39	0.726
ResNet101	38	0.728
ResNet101V2	51	0.737
ResNet152	107	**0.820**
ResNet152V2	15	0.690
VGG16	37	0.677
VGG19	70	0.715
Xception	22	0.730

macro average F1-score, precision, and recall, with improvements of 0.023, 0.029, 0.013 respectively. The reported AUC is slightly higher for the model trained on the full training set, however, this value is negligible with a difference of just 0.001.

Table 12. Macro average test performance metrics for the InceptionResNetV2 model trained on the full DFUC2021 training set and the InceptionResNetV2 model trained on the 80% similarity threshold training set. AUC - area under the curve.

Model	F1-Score	Precision	Recall	AUC
InceptionResNetV2 (full)	0.511	0.523	0.541	**0.841**
InceptionResNetV2 (80)	**0.534**	**0.552**	**0.554**	0.840

The F1-scores for the multi-class test performance of the InceptionResNetV2 model trained on the 80% similarity threshold training set are shown in Table 13. The InceptionResNetV2 model trained on the 80% similarity threshold training set shows a clear performance increase for all classes, with improvements of 0.011 for the none class, 0.025 for the infection class, 0.015 for the ischemia class,

Table 13. F1-score multi-class test results for the InceptionResNetV2 model trained on the full DFUC2021 training set and the InceptionResNetV2 model trained on the 80% similarity threshold training set.

Model	None	Infection	Ischemia	Both	Accuracy
Full	0.707	0.512	0.431	0.394	0.602
80%	**0.718**	**0.537**	**0.446**	**0.433**	**0.621**

and 0.039 for the both class (infection and ischemia). The biggest performance increase is shown for the both class (0.039). Accuracy for all classes is 0.602 for the model trained on the full DFUC2021 training set, and 0.621 for the model trained on the 80% similarity threshold training set. This demonstrates an accuracy improvement of 0.019 when testing using the InceptionResNetV2 model trained on the 80% similarity threshold training set.

We observe that a number of the visually similar images identified by the dupeGuru fuzzy algorithm were examples of natural augmentation. Our findings indicate that the excess use of subtle augmentation cases does not have the desired effect of boosting network performance. This highlights the importance of rigorously experimenting using individual augmentation sets when training deep learning networks to ascertain if models are being negatively affected by certain augmentation types. We encourage researchers working in other deep learning domains to follow these guidelines in future work to ensure that models are effectively trained, and that the effect of individual augmentation types is better understood.

Our experiments focused on the use of a single image similarity algorithm - an open-source fuzzy algorithm found in the dupeGuru application. Future research might test other image similarity methods, such as the structural similarity index measure, cosine similarity, or mean squared error [2].

6 Conclusion

In this work we observed and quantified the effect of non-identical similar images on a selection of popular deep learning multi-class classification networks trained using a large publicly available diabetic foot ulcer dataset. We found that model accuracy is negatively affected by the presence of non-identical visually similar images, but that the removal of too many non-identical visually similar images can degrade network performance. We report our findings to encourage researchers to experiment with other deep learning datasets to gauge a better understanding of the effect of image similarity and the potential bias it may introduce into models trained on such data.

Acknowledgment. We gratefully acknowledge the support of NVIDIA Corporation who provided access to GPU resources for the DFUC2020 and DFUC2021 Challenges.

References

1. Yap, M.H., Cassidy, B., Pappachan, J.M., O'Shea, C., Gillespie, D., Reeves, N.D.: Analysis towards classification of infection and ischaemia of diabetic foot ulcers. In: 2021 IEEE EMBS International Conference on Biomedical and Health Informatics (BHI), pp. 1–4. IEEE (2021)
2. Cassidy, B., Kendrick, C., Brodzicki, A., Jaworek-Korjakowska, J., Yap, M.H.: Analysis of the ISIC image datasets: usage, benchmarks and recommendations. Med. Image Anal. **75**, 102305 (2022)
3. Goyal, M., Reeves, N., Rajbhandari, S., Yap, M.H.: Robust methods for real-time diabetic foot ulcer detection and localization on mobile devices. IEEE J. Biomed. Health Inform. **23**, 1730–1741 (2018)
4. Cassidy, B., et al.: The DFUC 2020 dataset: analysis towards diabetic foot ulcer detection. touchREVIEWS Endocrinol. **17**, 5–11 (2021)
5. Yap, M.H., et al.: Deep learning in diabetic foot ulcers detection: a comprehensive evaluation. Comput. Biol. Med. **135**, 104596 (2021)
6. Reeves, N.D., Cassidy, B., Abbott, C.A., Yap, M.H.: Novel technologies for detection and prevention of diabetic foot ulcers, chapter 7. In: Gefen, A. (ed.) The Science, Etiology and Mechanobiology of Diabetes and its Complications, pp. 107–122. Academic Press (2021)
7. Cassidy, B., et al.: A cloud-based deep learning framework for remote detection of diabetic foot ulcers. IEEE Pervasive Comput. (01), 1–9 (2022)
8. Pappachan, J.M., Cassidy, B., Fernandez, C.J., Chandrabalan, V., Yap, M.H.: The role of artificial intelligence technology in the care of diabetic foot ulcers: the past, the present, and the future. World J. Diab. **13**, 1131–1139 (2022)
9. Goyal, M., Reeves, N.D., Rajbhandari, S., Ahmad, N., Wang, C., Yap, M.H.: Recognition of ischaemia and infection in diabetic foot ulcers: dataset and techniques. Comput. Biol. Med. **117**, 103616 (2020)
10. Cassidy, B., et al.: Diabetic foot ulcer grand challenge 2021: evaluation and summary. In: Yap, M.H., Cassidy, B., Kendrick, C. (eds.) DFUC 2021. LNCS, vol. 13183, pp. 90–105. Springer, Cham (2022). https://doi.org/10.1007/978-3-030-94907-5_7
11. Yap, M.H., Kendrick, C., Reeves, N.D., Goyal, M., Pappachan, J.M., Cassidy, B.: Development of diabetic foot ulcer datasets: an overview. In: Yap, M.H., Cassidy, B., Kendrick, C. (eds.) DFUC 2021. LNCS, vol. 13183, pp. 1–18. Springer, Cham (2022). https://doi.org/10.1007/978-3-030-94907-5_1
12. Goyal, M., Yap, M.H., Reeves, N.D., Rajbhandari, S., Spragg, J.: Fully convolutional networks for diabetic foot ulcer segmentation. In: 2017 IEEE International Conference on Systems, Man, and Cybernetics (SMC), pp. 618–623. IEEE (2017)
13. Kendrick, C., et al.: Translating clinical delineation of diabetic foot ulcers into machine interpretable segmentation (2022)
14. Goyal, M., Reeves, N.D., Davison, A.K., Rajbhandari, S., Spragg, J., Yap, M.H.: DFUNet: convolutional neural networks for diabetic foot ulcer classification. IEEE Trans. Emerg. Top. Comput. Intell. **4**(5), 728–739 (2018)
15. Al-Garaawi, N., Ebsim, R., Alharan, A.F.H., Yap, M.H.: Diabetic foot ulcer classification using mapped binary patterns and convolutional neural networks. Comput. Biol. Med. **140**, 105055 (2022)
16. dupeGuru (2018). https://dupeguru.voltaicideas.net/. Accessed 7 June 2022

DFUC2022 Challenge Papers

HarDNet-DFUS: Enhancing Backbone and Decoder of HarDNet-MSEG for Diabetic Foot Ulcer Image Segmentation

Ting-Yu Liao(✉) [ORCID], Ching-Hui Yang [ORCID], Yu-Wen Lo [ORCID], Kuan-Ying Lai [ORCID], Po-Huai Shen [ORCID], and Youn-Long Lin [ORCID]

Department of Computer Science, National Tsing Hua University, Hsinchu, Taiwan
{wendy107062324,hui09080729,wagw1014,kytimmylai}@gapp.nthu.edu.tw,
ylin@cs.nthu.edu.tw

Abstract. Diabetic foot ulcers are caused by neuropathic and vascular complications of diabetes mellitus. In order to provide a proper diagnosis and treatment, wound care professionals need to extract accurate morphological features from the foot wounds. Using computer-aided systems is a promising approach to extract related morphological features and segment the lesions. We propose a convolution neural network called HarDNet-DFUS by enhancing the backbone and replacing the decoder of HarDNet-MSEG, which was the state-of-the-art network for colonoscopy polyp segmentation in 2021. For the MICCAI 2022 Diabetic Foot Ulcer Segmentation Challenge (DFUC2022), we train HarDNet-DFUS using the DFUC2022 dataset and increase its robustness by means of five-fold cross validation and Test Time Augmentation. In the validation phase of DFUC2022, HarDNet-DFUS achieved 0.7063 mean Dice and was ranked third among all participants. In the final testing phase of DFUC2022, it achieved 0.7287 mean Dice and was the first place winner. The code is available on https://github.com/kytimmylai/DFUC2022.

Keywords: Medical imaging · Diabetic foot ulcer image segmentation · Deep learning · Neural network

1 Introduction

Diabetes is a global epidemic, and it is estimated that by the end of 2045, approximately 600 million people will have diabetes [3]. Diabetic Foot Ulcers (DFUs) is one of the complications of diabetes, often leading to more serious conditions such as infection and ischemia, which can significantly prolong treatment and often lead to amputation and, in more severe cases, death. In current practice,

---

T.-Y. Liao, C.-H. Yang, Y.-W. Lo and K.-Y. Lai—These authors contributed equally to this work.

medical professionals primarily use manual measurement tools to visually examine and evaluate patients with DFU to determine its severity. However, this is not only time consuming but also challenging for podiatrists. Therefore, speed and accuracy become important aspects to accurately determine the exact region of the ulcer.

In recent years, based on the rapid development of convolutional neural networks, many deep learning techniques have been applied in the field of medical imaging. For this task, U-Net [11] employs an encoder-decoder architecture that has achieved breakthrough performance and stimulated many improvements, such as ResUNet++ [8], DoubleU-Net [7], UNet++ [19], etc. However, the overly complex network architecture, low accuracy of small target segmentation, and slow segmentation speed have limited the practical deployment of U-Net variants in the clinical field.

Therefore, based on the previous state-of-the-art HarDNet-MSEG [6] for colonoscopy polyp segmentation, we enhance its backbone incorporating the concept of CSPNet [13] and ShuffleNetV2 [10], and employ a new decoder introduced in the Lawin Transformer [16]. We called the resultant network HarDNet-DFUS. It improves the capability of detecting ulcer regions and can deliver better accuracy compared to the original HarDNet-MSEG.

The contributions of this study can be summarized as follows: First, we have improved the HarDNet-MSEG model to achieve better performance in ulcer region segmentation. Second, we have enhanced the HarDNet [2] backbone to achieve higher speed. Third, we have evaluated the proposed method using the single-class segmentation tasks of the DFUC2022 Challenge.

2   Method

Figure 1 depicts the original HarDNet-MSEG and our enhanced model. Our enhancement includes modifying each HarDBlock module in the encoder backbone with a new HarDBlockV2 module and replacing Receptive Field Block (RFB) modules in the decoder with that of Lawin Transformer [16].

2.1   HarDNetV2 – Channel Balanced HarDNet

HarDNet-MSEG's backbone consists of basic building blocks called HarDBlock. Our enhanced backbone incorporates ideas from CSPNet and ShuffleNetV2, and we call it HarDBlockV2 as depicted in Fig. 2. To achieve the best MACs over CIO ratio (MoC) proposed by HarDNet [2], we perform channel splitting on the outputs of a convolutional layer l according to its number of output connections. This makes the number of input channels equal to the number of output channels for each Conv3x3 layer. According to the design principle of HarDNet, the amount of DRAM access could be reduced.

In addition, we introduce a new link pattern that simplifies the network architecture design. We build the inter-layer connections according to the factors of the desired block depth n. For example, when n = 9, its factors are 1, 3, and

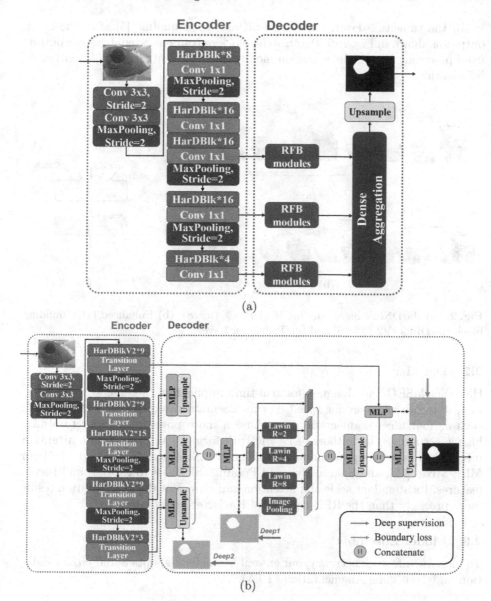

Fig. 1. (a) Original HarDNet-MSEG model, (b) Enhanced model by replacing HarD-Block with HarDBlockV2 and replacing RFB modules with the decoder of Lawin Transformer.

9, so we create shortcuts to the 1st, 3rd, and 9th convolutional layers. By doing so, the depth of a basic building block in HarDNetV2 is no longer constrained to the power of 2. Instead of 4, 8, or 16 employed by HarDBlock, we choose block depth $n = 3$, 9, and 15 to build HarDBlockV2, resulting in less data movement with the same number of convolutional layers.

In the transition layer, we add an SE attention module [5] after the block output as shown in Fig. 2(c). Because the block output concatenates some output from preceding layers, the attention module facilitates utilization of multi-scale information.

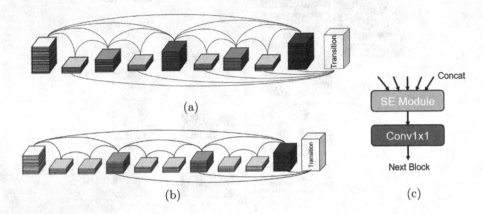

Fig. 2. (a) HarDNet's basic building HarDBlock (n = 8); (b) Enhanced basic building block HarDBlockV2 (n = 9); and (c) Transition Layer.

2.2   Decoder

HarDNet-MSEG was designed for real-time application of colonoscopy. Therefore, it trades accuracy for speed. For an accuracy-oriented non-real-time task such as foot ulcer segmentation, we chose a more powerful decoder to obtain higher accuracy. The authors of Lawin Transformer [16] proposed an attention mechanism called Large Window Attention. It utilizes an MLP Decoder [15], an MLP-Mixer [12], and Spatial Pyramid Pooling (SPP) [4] to capture multi-scale features. Its abundant scale and attention can represent the segmentation result more precisely than the RFB decoder of HarDNet-MSEG does.

2.3   Model Ensemble

To increase inference accuracy, our ensemble strategy adopts 5-fold cross validation and Test Time Augmentation (TTA).

5-Fold Cross Validation. The dataset is randomly partitioned into five folds of 400 images each. For each cross-fold iteration, four folds are used for training and the fifth for validation. After five iterations, we obtain five derived submodels.

Test Time Augmentation. During inference, we perform TTA on each submodel. That is, to generate an additional image via image flipping, feed both the test images and the additional image to these sub-models, and then take the average of their outputs as our prediction results.

2.4 Loss Function

Our loss function for DFUC2022 segmentation is given in Eq. (1), which calculates the loss between the ground truth G, the output O of our model, the output D_i of the deep supervision and the output B of the boundary.

$$L = l(G, O) + \sum_i l(G, D_i) + l_{BCE}(G_B, B) \qquad (1)$$

where $l(G, O) = l_{BCE}^w(G, O) + l_{IoU}^w(G, O)$, and l_{IoU}^w and l_{BCE}^w denote weighted IoU loss and weighted BCE loss, respectively. These two functions have the same definition as that of [14]. $l_{BCE}(G_B, B)$ calculates the loss between the prediction boundary and the ground truth boundary.

2.5 Post-processing

We pass the output through the Tanh function and normalize the result into the range $[0, 1]$ and round to $\{0, 1\}$ to represent a mask. The last step is hole filling. We first flood-fill the mask prediction from point $(0, 0)$, then invert it as an inverted mask. Finally, we apply a logical OR operation to the original mask to invert it to obtain the hole-filled image as our final mask image.

3 Experiments

3.1 Settings

We train the proposed models on a single NVIDIA Tesla V100 GPU. The batch size is 6 and learning rate is 1e−4 with cosine annealing schedule. Training the model for 300 epochs takes about 15 h. To keep their original aspect ratio, the training images are zero-padded into squares and then resized to 512×512. We also employed multi-scaling. The image would be randomly resized into multiples of 64 between 384 (512×0.75) and 640 (512×1.25).

Data augmentation includes random vertical flipping, horizontal flipping, cropping, shifting, scaling, rotation, coarse dropout, brightness changing, contrast changing, and Gaussian noise introduction.

Our measurement metric is Dice coefficient (Dice), Jaccard Index (Jaccard), False Positive Error (FPE), and False Negative Error (FNE) [9]. The evaluation results are our submissions during the validation phase of DFUC2022.

3.2 Dataset

The DFUC2022 dataset [9] is provided by the organizer of the MICCAI 2022 Diabetic Foot Ulcer Challenge [18]. We refer the readers to the overview and detailed description of the creation of diabetic foot ulcer datasets by Yap et al. [17]. The dataset comprises 2000 training images and 2000 testing images with single-class ulcer segmentation labels. The size of all images is 640×480 as described in preparation of the DFUC2020 dataset [1]. Regarding ulcer sizes, 89% of the ulcer areas are smaller than 5% of the total image size.

3.3   Experiment Results

First, we compare our enhanced backbone HarDNetV2 with the original HarD-
Net. Table 1 shows that our new backbone gained 1% accuracy while keeping a
similar speed to the original HarDNet.

Table 1. Effectiveness of new backbone

Model	Dice	Jaccard	FPE	FNE	FPS
HarDNet-MSEG	0.6553	0.5522	**0.2728**	0.2949	**108**
HarDNetV2-53-RFB	**0.6651**	**0.5620**	0.2788	**0.2731**	104

Table 2 shows the mean Dice improvement after using 5-fold cross validation.
5-fold ensemble provides a 0.8% accuracy gain.

Table 2. Effectiveness of 5-fold cross-validation and ensemble

Model	5-Fold	Dice	Jaccard	FPE	FNE
HarDNetV2-53-RFB		0.6651	0.5620	0.2788	0.2731
	✓	**0.6730**	**0.5698**	**0.2716**	**0.2695**

In Table 3, we further compare different combinations of backbones
(HarDNetV2-53 and HarDNetV2-CSP69) and decoders (RFB module and Lawin
decoder). We designate the best architecture HarDNet-DFUS, i.e., the one with
HarDNetV2 backbone (53 convolution layers) and Lawin decoder.

Table 3. Results of different combinations of new backbone sizes and decoder types

Model	5-Fold	Dice	Jaccard	FPE	FNE
HarDNetV2-53-RFB	✓	0.6730	0.5698	0.2716	0.2695
HarDNetV2-CSP69-RFB	✓	0.6842	0.5801	0.2571	0.2702
HarDNetV2-53-Lawin (HarDNet-DFUS)	✓	**0.6950**	**0.5926**	**0.2413**	**0.2593**
HarDNetV2-CSP69-Lawin	✓	0.6870	0.5850	0.2539	0.2687

Figure 3 shows the loss and Dice of HarDNet-DFUS during training in one
of five folds. We plot the loss, deep supervision loss (deep1, deep2), boundary
loss (boundary loss), mean Dice (dice), the current best Dice (best dice) and the
validation loss (val_loss) at each epoch.

As shown in Table 4 and Table 5, we experiment with different deep supervision and TTA methods. There are two deep supervision losses and a boundary

loss. We can see that deep supervision loss works when we take more than one to join the training. TTA includes none, horizontal flip (hflip), vertical flip (vflip), and horizontal flip with a vertical flip (vhflip). These methods increase the accuracy in some cases, however, the effect is not robust.

We observe some small values being classified as positive after being compressed by the Sigmoid function, but not by the Tanh function. Therefore, we compare Sigmoid and Tanh and show the results in Table 6. Generally, Tanh provides better results.

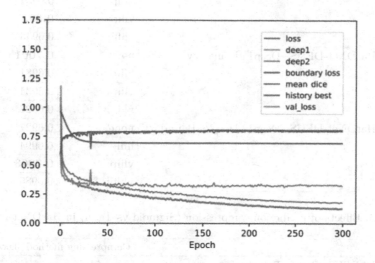

Fig. 3. Training process of HarDNet-DFUS (HarDNetV2-53-Lawin) in one of five folds.

Table 4. Segmentation accuracy of HarDNet-DFUS using different combinations of loss functions

Deep1	Deep2	Boundary	Dice	Jaccard	FPE	FNE
			0.6915	0.5876	0.2520	0.2598
✓			0.6852	0.5827	0.2524	0.2641
✓	✓		0.6950	0.5926	0.2413	0.2593
✓		✓	**0.7001**	**0.5981**	**0.2404**	**0.2540**
✓	✓	✓	0.6927	0.5901	0.2423	0.2651

For the validation phase of DFUC2022, HarDNet-DFUS achieved 0.7063 mean Dice and ranked third among 21 participating teams. During the final testing phase, HarDNet-DFUS achieved 0.7287 mean Dice and ranked first among all teams. We also tried using the polyp-oriented HarDNet-MSEG for the DFUC task, which resulted in a mean Dice reduction of 5%.

Table 5. Effect of different Test Time Augmentations in HarDNet-DFUS

Model	TTA Method	Dice
HarDNet-DFUS	none	0.6920
	hflip	0.6947
	vflip	0.6931
	vhflip	**0.6975**
HarDNet-DFUS+Deep1+Deep2	none	0.6950
	hflip	0.6958
	vflip	**0.6992**
	vhflip	0.6943
HarDNet-DFUS+Deep1+Boundary	none	**0.7001**
	hflip	0.6981
	vflip	0.6934
	vhflip	0.6928
HarDNet-DFUS+Deep1+Deep2+Boundary	none	0.6927
	hflip	0.6994
	vflip	**0.7063**
	vhflip	0.6985

Table 6. Effects of prediction compression (Sigmoid vs Tanh) in HarDNet-DFUS

Model	Compressing method	Dice
HarDNet-DFUS+Deep1+Boundary	Sigmoid	0.6752
	Tanh	**0.7001**
HarDNet-DFUS+Deep1+Boundary (w/hflip)	Sigmoid	0.6834
	Tanh	**0.6981**
HarDNet-DFUS+Deep1+Deep2+Boundary (w/hflip)	Sigmoid	0.6950
	Tanh	**0.6994**
HarDNet-DFUS+Deep1+Deep2+Boundary (w/vflip)	Sigmoid	0.7029
	Tanh	**0.7063**
HarDNet-DFUS+Deep1+Deep2+Boundary (w/vhflip)	Sigmoid	**0.6995**
	Tanh	0.6985

4   Conclusion and Future Work

For the task of diabetic foot ulcer segmentation, we have proposed enhancing the previous state-of-the-art HarDNet-MSEG polyp segmentation network with a new backbone and a more effective decoder. We call the new network HarDNet-DFUS. Five-fold cross validation, deep supervision, boundary supervision, and test time augmentation together contribute to an approximate improvement of 5% in mean Dice compared with the original HarDNet-MSEG. We have participated in the 2022 MICCAI DFUC Challenge and have been awarded the first place winner.

In the future, we would like to deploy HarDNet-DFUS in clinical fields and expand the application scope of this work to other medical imaging tasks.

Acknowledgements. This research is partially supported by the Ministry of Science and Technology (MOST) of Taiwan. We thank the National Center for High-performance Computing (NCHC) for computational and storage resources. We would also like to thank Professor Tzu-Chen Dorothy Yen and Professor Chang-Fu Kuo of Chang-Gang Memorial Hospital for their advice.

References

1. Cassidy, B., et al.: The DFUC 2020 dataset: analysis towards diabetic foot ulcer detection. touchREVIEWS in Endocrinol. **17**, 5–11 (2021). https://doi.org/10.17925/EE.2021.17.1.5. https://www.touchendocrinology.com/diabetes/journal-articles/the-dfuc-2020-dataset-analysis-towards-diabetic-foot-ulcer-detection/1
2. Chao, P., Kao, C.Y., Ruan, Y.S., Huang, C.H., Lin, Y.L.: HarDNet: a low memory traffic network. In: Proceedings of the IEEE International Conference on Computer Vision (ICCV), pp. 3552–3561 (2019). https://doi.org/10.1109/ICCV.2019.00365
3. Cho, N., et al.: IDF diabetes atlas: global estimates of diabetes prevalence for 2017 and projections for 2045. Diab. Res. Clin. Pract. **138**, 271–281 (2018). https://doi.org/10.1016/j.diabres.2018.02.023
4. He, K., Zhang, X., Ren, S., Sun, J.: Spatial pyramid pooling in deep convolutional networks for visual recognition. IEEE Trans. Pattern Anal. Mach. Intell. **37**, 1904–1916 (2015). https://doi.org/10.1109/TPAMI.2015.2389824
5. Hu, J., Shen, L., Sun, G.: Squeeze-and-excitation networks. In: 2018 IEEE/CVF Conference on Computer Vision and Pattern Recognition, pp. 7132–7141 (2018). https://doi.org/10.1109/CVPR.2018.00745
6. Huang, C.H., Wu, H.Y., Lin, Y.L.: HarDNet-MSEG: a simple encoder-decoder polyp segmentation neural network that achieves over 0.9 Mean Dice and 86 FPS. arXiv preprint arXiv:2101.07172 (2021)
7. Jha, D., Riegler, M.A., Johansen, D., Halvorsen, P., Johansen, H.: DoubleU-Net: a deep convolutional neural network for medical image segmentation. In: 2020 IEEE 33rd International Symposium on Computer-Based Medical Systems (CBMS), pp. 558–564 (2020). https://doi.org/10.1109/CBMS49503.2020.00111
8. Jha, D., Smedsrud, P.H., Riegler, M.A., et al.: ResUNet++: an advanced architecture for medical image segmentation. In: 2019 IEEE International Symposium on Multimedia (ISM), pp. 225–2255 (2019). https://doi.org/10.1109/ISM46123.2019.00049
9. Kendrick, C., et al.: Translating clinical delineation of diabetic foot ulcers into machine interpretable segmentation (2022). https://doi.org/10.48550/ARXIV.2204.11618
10. Ma, N., Zhang, X., Zheng, H.T., Sun, J.: ShuffleNet V2: practical guidelines for efficient CNN architecture design. In: Proceedings of the European Conference on Computer Vision (ECCV) (2018)
11. Ronneberger, O., Fischer, P., Brox, T.: U-Net: convolutional networks for biomedical image segmentation. In: Navab, N., Hornegger, J., Wells, W.M., Frangi, A.F. (eds.) MICCAI 2015. LNCS, vol. 9351, pp. 234–241. Springer, Cham (2015). https://doi.org/10.1007/978-3-319-24574-4_28

12. Tolstikhin, I.O., et al.: MLP-mixer: an all-MLP architecture for vision. In: Advances in Neural Information Processing Systems, vol. 34, pp. 24261–24272 (2021). https://doi.org/10.48550/ARXIV.2105.01601

13. Wang, C.Y., Mark Liao, H.Y., Wu, Y.H., Chen, P.Y., Hsieh, J.W., Yeh, I.H.: CSPNet: a new backbone that can enhance learning capability of CNN. In: 2020 IEEE/CVF Conference on Computer Vision and Pattern Recognition Workshops (CVPRW), pp. 1571–1580 (2020). https://doi.org/10.1109/CVPRW50498.2020.00203

14. Wei, J., Wang, S., Huang, Q.: F^{3}Net: fusion, feedback and focus for salient object detection. In: Proceedings of the AAAI Conference on Artificial Intelligence (2019). https://doi.org/10.48550/ARXIV.1911.11445

15. Xie, E., Wang, W., Yu, Z., Anandkumar, A., Alvarez, J.M., Luo, P.: SegFormer: simple and efficient design for semantic segmentation with transformers. In: Advances in Neural Information Processing Systems, vol. 34, pp. 12077–12090 (2021). https://doi.org/10.48550/ARXIV.2105.15203

16. Yan, H., Zhang, C., Wu, M.: Lawin transformer: improving semantic segmentation transformer with multi-scale representations via large window attention. arXiv preprint arXiv:2201.01615 (2022)

17. Yap, M.H., Kendrick, C., Reeves, N.D., Goyal, M., Pappachan, J.M., Cassidy, B.: Development of diabetic foot ulcer datasets: an overview. Diab. Foot Ulcers Grand Challenge 1–18 (2021)

18. Yap, M.H., et al.: Diabetic foot ulcers grand challenge 2022 (2021). https://doi.org/10.5281/zenodo.6389665

19. Zhou, Z., Rahman Siddiquee, M.M., Tajbakhsh, N., Liang, J.: UNet++: a nested U-Net architecture for medical image segmentation. In: Stoyanov, D., et al. (eds.) DLMIA/ML-CDS -2018. LNCS, vol. 11045, pp. 3–11. Springer, Cham (2018). https://doi.org/10.1007/978-3-030-00889-5_1

OCRNet for Diabetic Foot Ulcer Segmentation Combined with Edge Loss

Huahui Yi[1,4], Wei Xu[1,2,4], Zekun Jiang[1,4], Jun Gao[1,4], Qingbo Kang[1,4], Qicheng Lao[3,4,5(✉)], and Kang Li[1,4,5(✉)]

[1] West China Biomedical Big Data Center, West China Hospital, Sichuan University, Chengdu 610041, Sichuan, People's Republic of China
[2] Suzhou Institute for Advanced Research, University of Science and Technology of China, Suzhou 215123, People's Republic of China
[3] School of Artificial Intelligence, Beijing University of Posts and Telecommunications, Beijing 100876, People's Republic of China
qicheng.lao@bupt.edu.cn
[4] West China Hospital-SenseTime Joint Lab, Chengdu 610041, Sichuan, People's Republic of China
likang@wchscu.cn
[5] Shanghai Artificial Intelligence Laboratory, Shanghai 200232, People's Republic of China

Abstract. Diabetic foot ulcer is a serious manifestation of lesions on the diabetic foot that requires close monitoring and management. The research at hand investigates an approach on segmentation of diabetic foot ulcer area, conducted as part of the Diabetic Foot Ulcer Challenge (DFUC) 2022. We use OCRNet as the baseline for segmentation and a powerful ConvNeXt network was adopted as the backbone. To obtain better results, a boundary loss was introduced to further constrain the boundary of segmentation. In addition, gamma correction was used in the inference stage in order to reduce the difference in luminance between the training, validation and test sets. Our method won 2nd place in the DFUC2022 with a Dice score of 72.80%. Source code is available at: DFUC2022SegmentationOcrnet.

Keywords: Diabetic foot ulcer · Medical segmentation · Deep learning

1 Introduction

Diabetic foot (DF) disease is one of the common serious chronic complications of diabetes, and it is also the most common cause of non-traumatic amputation. Diabetic foot disease not only reduces the quality of life of patients, affecting their physical and mental health, but also greatly increases the economic pressure of patients. Studies have shown that about 25% of people with diabetes will develop foot ulcers of varying degrees [1]. Due to the influence of a variety of risk factors, diabetic foot is clinically manifested in the majority of elderly patients, and the clinical symptoms are not the same, mostly manifested as diabetic foot ulcers (DFU).

© The Author(s), under exclusive license to Springer Nature Switzerland AG 2023
M. H. Yap et al. (Eds.): DFUC 2022, LNCS 13797, pp. 31–39, 2023.
https://doi.org/10.1007/978-3-031-26354-5_3

In order to carry out proper treatment, it is usually necessary to measure the shape characteristics of diseased wounds. In addition, the wound area is an important evaluation metric of the final treatment result, and the healing process can be evaluated by monitoring the wound area, so as to evaluate the treatment effect [6]. In clinical practice, the morphological characteristics of foot ulcers are usually studied by medical experts through visual observation. However, this observation method is very subjective, error-prone and time-consuming [7]. In recent research, there have been many studies that focus on automated segmentation of DFU from DFU wound images, and they have provided important insights into the development of deep neural networks in this domain [2,11].

In this paper, we proposed a ConvNeXt [9] backboned OCRNet [14] to segment the DFU with a coarse to fine training manner, and introduced an Edge BCE loss to constrain boundary information for an improved segmentation result. The final results on DFUC2022 was promising as we obtained a Dice score of 72.80% on the test dataset which ranked 2nd in the challenge.

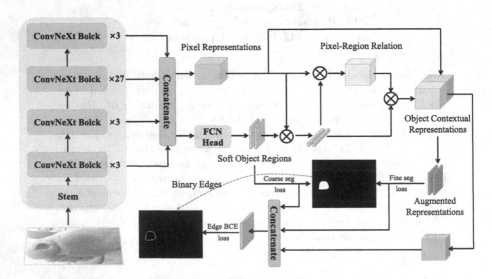

Fig. 1. Structure of our Edge OCRNet for DFUC2022 segmentation. We replaced the original backbone HRNet [15] with the simple and powerful ConvNeXt [9] and an Edge BCE loss was introduced to obtain an improved segmentation result.

2   Methodology

The framework of our method was shown in Fig. 1, where a ConvNeXt [9] backboned OCRNet [14] was used to get the diabetic foot ulcers segmentation with a coarse to fine manner, and an Edge BCE loss was introduced to obtain a better segmentation result.

2.1 Datasets

DFUC2022: The DFUC2022 dataset[1] is from the MICCAI DFU Grand Challenge 2022 [13] and it is the largest DFU segmentation dataset with ground truth delineation, which consists of a training set of 2000 images and a testing set of 2000 images [8]. Figure 2 shows some example images from the DFUC2022 datasets, where the images were originally prepared in DFUC2020 [3]. The DFUC2022 training set consists of 2304 ulcers and nearly 90% (2054 out of 2304) of the ulcers are less than 5% of the total image size [8]. This represents a significant challenge for segmentation algorithms as small regions are hard to delineate by deep learning algorithms.

ImageNet: The ImageNet [5] dataset is a large-scale labeled natural image dataset of approximately 15 million images organized according to the WordNet [12] architecture. Compared with other medical image datasets, DFUC2022 is closer to natural images, so we pre-trained on ImageNet and trained the segmentation model based on the pre-trained model.

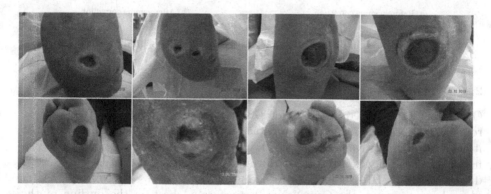

Fig. 2. Image samples from the DFUC2022 dataset [8], the similar images created during DFUC2020 [3].

Gamma Correction. Gamma correction is a nonlinear operation used to encode and decode luminance or tristimulus values in video or still image systems which compensates for the loss of brightness by modifying the gamma value. Gamma correction is, in the simplest cases, defined by the following power-law expression:

$$V_{out} = AV_{in}^{1/\gamma} \tag{1}$$

where a gamma value $\gamma > 1$ is sometimes called an encoding gamma, and the process of encoding with this compressive power-law nonlinearity is called gamma compression; conversely a gamma value $\gamma < 1$ is called a decoding

---
[1] https://dfu-challenge.github.io/.

gamma, and the application of the expansive power-law nonlinearity is called gamma expansion. As shown in Fig. 3, we set a gamma value of 1.1 to apply gamma correction on the whole dataset and it can be observed that the difference between the mean RGB values of the validation and test sets and the training set shrinks after using the gamma correction. In our practice, OCRNet using swin-transformer as the backbone performs best when tested with a gamma value of 1.08, whereas using ConvNeXt as the backbone, a larger gamma value of 1.10 can achieve better performance. The experiment results are shown in Table 3.

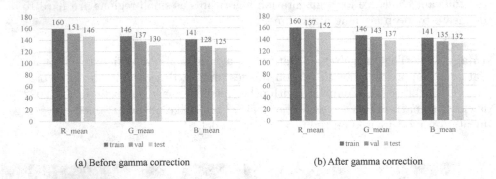

(a) Before gamma correction   (b) After gamma correction

Fig. 3. Differences in dataset distribution before and after gamma correction.

2.2   Network

Our approach is based on OCRNet [14] which can aggregate better context representation in the semantic segmentation task, and it was demonstrated that it achieved competitive performance on various benchmarks. We use a variety of networks as backbones to test the performance of OCRNet [14] on the DFUC2022 dataset, such as HR-Net [15], Swin Transforme [10] and ConvNeXt [9]. After several experiments, the ConvNeXt [9] performed best as the outline of the diabetic foot ulcer wound is relatively clear, therefore this ConvNeXt [9] was chosen as the backbone of our OCRNet model.

The overall framework of our method is shown in Fig. 1. Our method uses OCRNet [14] to construct contextual information as it can enhance the contribution of pixels from the same class of objects, and the idea starts from the definition of semantic segmentation and is consistent with first principles thinking. To obtain better performance, we replaced the original backbone HRNet [15] with ConvNeXt [9] as it is the latest powerful standard convolution network while retaining simplicity and efficiency. To obtain a better segmentation of the diabetic foot ulcer wound, we add a loss function named Edge loss to constrain the boundary information. We generate the boundary maps from the segmentation masks and use standard Binary Cross-Entropy (BCE) loss on the predicted boundary maps B and the generated ground truth boundary maps $\hat{B}$:

$$L_{Edge} = L_{BCE}(B, \hat{B}) \tag{2}$$

The whole loss is a weighted sum of Coarse-Seg loss L_{Coarse}, Fine-Seg loss L_{Fine} and Edge BCE loss L_{Edge}:

$$Loss = \alpha L_{Coarse} + \beta L_{Fine} + \theta L_{Edge} \qquad (3)$$

where we set $\alpha = 0.4$, $\beta = 1.0$, and $\theta = 20$ in our experiments. The Coarse-Seg loss L_{Coarse} and Fine-Seg loss L_{Fine} are both Cross Entropy Loss.

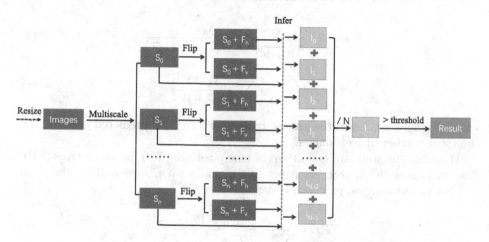

Fig. 4. The processing flow of our model during the prediction phase.

2.3   Training

During training, we use AdamW as the optimizer, set the initial learning rate as 0.00008, weight decay as 0.05, warmup steps as 1500, warmup ratio as 1e−6, batch size as 4, and training iterations as 60K if not specified. We use horizontal flipping (with 0.5 probability), random cropping (cropping size 512×512, maximum crop rate is 0.75), multi-scale training (image size is from 0.5 to 2.0 with stride 0.25), and Photometric Distortion [4] as augmentation techniques.

2.4   Post-processing

In the test phase, we first resize the image to 576×768 pixels, then adjusted the lightness and darkness of the image using Gamma Correction (the gamma value defaults to 1.11), then used a Multi-Scale testing and Test Time Augmentation (horizontal flip and vertical flip) as shown in Fig. 4. The results were obtained by testing all the images for the enhancements, and then all the results were summed and averaged. When the predicted score is greater than the threshold, it is considered as a diabetic foot ulcer (foreground), here the threshold value is 0.655 by default.

2.5   Evaluation Metrics

To evaluate the segmentation performance of our proposed method we use the Dice Similarity Index (Dice) (Eq. 4), mean Intersection Over Union (mIoU) (Eq. 5), False Positive Error (FPE) (Eq. 6) and False Negative Error (FNE) (Eq. 7) as suggested in [8]. For the DFUC2022 challenge, we report the Dice, mIoU, FPE and FNE as these scores were provided by the challenge organisers (the Dice score was used as the main evaluation index in the challenge).

$$Dice = 2 * \frac{1}{n} * \sum_{i=0}^{n-1} \frac{|X_i \cap Y_i|}{|X_i| + |Y_i| - |X_i \cap Y_i|} \tag{4}$$

$$mIoU = \frac{1}{n} * \sum_{i=0}^{n-1} \frac{|X_i \cap Y_i|}{|X_i| + |Y_i|} \tag{5}$$

where X and Y represent the ground truth mask and the predicted mask, and n the total number of test images.

Based on the standard definitions of true positives (TP), false positives (FP), false negatives (FN), true negatives (TN), and with n representing the total number of test images, precision is defined as:

$$FPE = \frac{\sum_{i=0}^{n-1} FP}{\sum_{i=0}^{n-1} PF + \sum_{i=0}^{n-1} TN} \tag{6}$$

$$FNE = \frac{\sum_{i=0}^{n-1} FN}{\sum_{i=0}^{n-1} TN + \sum_{i=0}^{n-1} FN} \tag{7}$$

3   Experiments

In Table 1, in the case of the same network structure (OCRNet), replacing different backbone networks, as well as using the same backbone network but of different sizes, has a significant impact on the performance of the model. In our experiments, HRNet < Swin-Transformer < ConvNeXt, with larger backbone networks giving better results. Our best results were obtained by pre-training ConvNeXt-XLarge on ImageNet-21K. On this basis, we introduced Edge branches and Edge loss to enhance the ability of the model to detect wound edge information of diabetic foot ulcers and improve the overall performance on wound segmentation.

Table 1. The Dice score (%), mIoU (%) FPE (%) and FNE (%) of the diabetic foot ulcer segmentation under different networks and different strategies.

Model	Backbone	Dice	mIoU	FPE	FNE
OCRNet	HRNet-48	70.57	60.28	25.42	22.47
	Swin-T	71.20	60.93	25.11	21.12
	Swin-L	72.12	61.95	24.60	20.63
	ConvNeXt-B	71.73	61.51	25.79	19.72
	ConvNeXt-XL	72.19	61.94	25.48	**19.34**
Edge-OCRNet	ConvNeXt-XL	**72.26**	**62.07**	**24.75**	20.10

Table 2. The final results of DFUC2022.

Team	Approach	Dice	mIoU	FPE	FNE
yllab	HarDNet-DFUS	**72.87**	62.52	23.41	20.48
LkRobotAI Lab (Ours)	Edge-OCRNet	72.80	**62.76**	22.61	21.54
agaldran	–	72.63	62.73	**22.10**	22.62
ADAR-LAB	–	72.54	62.45	25.82	**18.47**
seoyoung	–	72.20	62.08	25.84	19.25
FHDO	–	71.69	61.30	24.53	21.45

With the presented methodology, we attended the DFUC2022 Challenge. Based on the single-scale optimal model, by using a combination of gamma correction, multiple scale enhancements and TTA at test time, we achieved second place on the DFUC2022 test set with a Dice of 72.80 (2nd), mIoU of 62.76 (1st), FPE of 22.61 (1st) and FNE of 21.54. More detailed results can be found in Table 2. Figure 5 shows four examples of segmentation results obtained by our method.

In addition, we have further explored the role of gamma correction on the DFUC2022 dataset. On the validation set (Table 3), scaling up the value of gamma compared to the original image (gamma = 1) improved the performance of the model, e.g. for gamma = 1.10, the model performance is improved by 0.12% in dice and 0.15% in mIoU compared to the original.

Table 3. Quantitative evaluation results of different gamma values on the validation set, where gamma of 1 means no gamma correction.

Gamma	Backbone	Dice	mIoU	FPE	FNE
1	ConvNeXt-XL	71.42	61.70	24.16	**23.45**
1.08	ConvNeXt-XL	71.49	61.78	23.58	23.94
1.10	ConvNeXt-XL	**71.54**	**61.85**	**23.32**	24.10

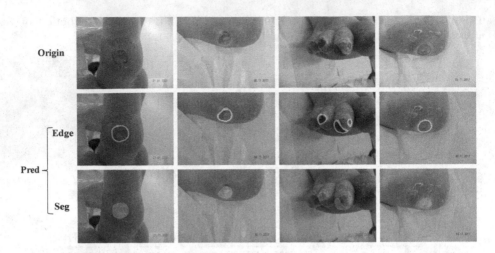

Fig. 5. Example results of our diabetic foot ulcer segmentation method.

4   Conclusions

In this work, we proposed a ConvNeXt [9] backboned OCRNet [14] with a coarse to fine training manner for DFU segmentation, and an Edge BCE loss was introduced to constrain the boundary information for a better segmentation result. We won 2nd place in DFUC2022 and a Dice score of 72.80% was obtained on the DFUC2022 test set. This study demonstrated the effectiveness of our Edge OCRNet for DFU segmentation, and the promising results lead us to believe that an efficient automated solution for DFU analysis and wound area measurement will be proposed in the future.

Acknowledgement. This study was supported by National Key Research and Development Program of China (2020YFB1711500, 2020YFB1711503), the 1·3·5 project for disciplines of excellence, West China Hospital, Sichuan University (ZYYC21004).

References

1. Baig, M.S., et al.: An overview of diabetic foot ulcers and associated problems with special emphasis on treatments with antimicrobials. Life **12**(7), 1054 (2022)
2. Bouallal, D., Douzi, H., Harba, R.: Diabetic foot thermal image segmentation using double encoder-ResUnet (DE-ResUnet). J. Med. Eng. Technol. 1–15 (2022). Taylor & Francis
3. Cassidy, B., et al.: The DFUC 2020 dataset: analysis towards diabetic foot ulcer detection. touchREVIEWS Endocrinol. **17**, 5–11 (2021). https://www.touchendocrinology.com/diabetes/journal-articles/the-dfuc-2020-dataset-analysis-towards-diabetic-foot-ulcer-detection/1
4. Contributors, M.: MMSegmentation: Openmmlab semantic segmentation toolbox and benchmark (2020). https://github.com/open-mmlab/mmsegmentation

5. Deng, J., Dong, W., Socher, R., Li, L.J., Li, K., Fei-Fei, L.: ImageNet: a large-scale hierarchical image database. In: 2009 IEEE Conference on Computer Vision and Pattern Recognition, pp. 248–255 (2009)

6. Goyal, M., Yap, M.H., Reeves, N.D., Rajbhandari, S., Spragg, J.: Fully convolutional networks for diabetic foot ulcer segmentation. In: 2017 IEEE International Conference on Systems, Man, and Cybernetics (SMC), pp. 618–623. IEEE (2017)

7. Jawahar, M., Anbarasi, L.J., Jasmine, S.G., Narendra, M.: Diabetic foot ulcer segmentation using color space models. In: 2020 5th International Conference on Communication and Electronics Systems (ICCES), pp. 742–747. IEEE (2020)

8. Kendrick, C., et al.: Translating clinical delineation of diabetic foot ulcers into machine interpretable segmentation. arXiv preprint arXiv:2204.11618 (2022)

9. Li, J., Wang, C., Huang, B., Zhou, Z.: ConvNeXt-backbone HoVerNet for nuclei segmentation and classification. arXiv preprint arXiv:2202.13560 (2022)

10. Liu, Z., et al.: Swin transformer: hierarchical vision transformer using shifted windows. In: Proceedings of the IEEE/CVF International Conference on Computer Vision, pp. 10012–10022 (2021)

11. Mahbod, A., Ecker, R., Ellinger, I.: Automatic foot ulcer segmentation using an ensemble of convolutional neural networks. arXiv preprint arXiv:2109.01408 (2021)

12. Miller, G.A.: WordNet: a lexical database for English. Commun. ACM **38**(11), 39–41 (1995)

13. Yap, M.H., et al.: Diabetic foot ulcers grand challenge 2022 (2021). https://doi.org/10.5281/zenodo.6389665

14. Yuan, Y., Chen, X., Chen, X., Wang, J.: Segmentation transformer: object-contextual representations for semantic segmentation. arXiv preprint arXiv:1909.11065 (2019)

15. Yuan, Y., Chen, X., Wang, J.: Object-contextual representations for semantic segmentation. In: Vedaldi, A., Bischof, H., Brox, T., Frahm, J.-M. (eds.) ECCV 2020. LNCS, vol. 12351, pp. 173–190. Springer, Cham (2020). https://doi.org/10.1007/978-3-030-58539-6_11

On the Optimal Combination of Cross-Entropy and Soft Dice Losses for Lesion Segmentation with Out-of-Distribution Robustness

Adrian Galdran[1,2]([✉]) [iD], Gustavo Carneiro[2] [iD],
and Miguel A. González Ballester[1,3] [iD]

[1] BCN Medtech, Department of Information and Communication Technologies,
Universitat Pompeu Fabra, Barcelona, Spain
{adrian.galdran,ma.gonzalez}@upf.edu
[2] University of Adelaide, Adelaide, Australia
gustavo.carneiro@adelaide.edu
[3] Catalan Institution for Research and Advanced Studies (ICREA), Barcelona, Spain

Abstract. We study the impact of different loss functions on lesion segmentation from medical images. Although the Cross-Entropy (CE) loss is the most popular option when dealing with natural images, for biomedical image segmentation the soft Dice loss is often preferred due to its ability to handle imbalanced scenarios. On the other hand, the combination of both functions has also been successfully applied in these types of tasks. A much less studied problem is the generalization ability of all these losses in the presence of Out-of-Distribution (OoD) data. This refers to samples appearing in test time that are drawn from a different distribution than training images. In our case, we train our models on images that always contain lesions, but in test time we also have lesion-free samples. We analyze the impact of the minimization of different loss functions on in-distribution performance, but also its ability to generalize to OoD data, via comprehensive experiments on polyp segmentation from endoscopic images and ulcer segmentation from diabetic feet images. Our findings are surprising: CE-Dice loss combinations that excel in segmenting in-distribution images have a poor performance when dealing with OoD data, which leads us to recommend the adoption of the CE loss for these types of problems, due to its robustness and ability to generalize to OoD samples. Code associated to our experiments can be found at https://github.com/agaldran/lesion_losses_ood.

Keywords: Lesion segmentation · Out-of-distribution generalization

1 Introduction

In the field of medical image segmentation, lesion/background separation is a central image analysis task that shows up in many different applications, ranging from polyp detection from endoscopic frames [1,14] to wound segmentation

M. H. Yap et al. (Eds.): DFUC 2022, LNCS 13797, pp. 40–51, 2023.
https://doi.org/10.1007/978-3-031-26354-5_4

[10,27] or volumetric brain tumor segmentation [15]. In all cases, the common setup is that we aim at partitioning the image domain into a foreground and a background region, with the foreground being typically smaller in size than the background. This results in a class-imbalanced segmentation problem for which model optimization becomes harder than in a well-balanced scenario.

A key aspect of optimization in imbalanced problems is the choice of the loss function [20]. In natural image segmentation tasks the standard Cross-Entropy (CE) loss is the most commonly adopted cost function [19], whereas in medical image segmentation the soft Dice loss enjoys widespread popularity [3]. The most likely reason for this is the typical imbalanced class distribution present in medical image segmentation: it has been shown that, for these problems, minimization of the soft Dice loss results in more accurate models, although it can induce potentially worse calibration [21]. Another widely adopted strategy is the combination of both CE and soft dice into a single loss function [20]. However, how to best combine each of them to achieve the reliability of CE *plus* the ability of soft Dice to handle class-imbalanced problems remains unclear.

Calibration is the property by which model predictions reflect true probabilities, and thereby meaningful uncertainties, as opposed to over or underconfident outputs. Hence, model calibration is a critical aspect of performance for sensitive use-cases like fine localization of tumor borders [21,24], but it is also important when dealing in test time with data that does not resemble the images used for model training. This sort of data is often referred to as Out-of-Distribution (OoD) [9,12], and it poses the challenge of how to learn models that can properly generalize to it. In this paper, we consider a specific but very relevant instance of an OoD generalization problem, namely models trained on images that always contain a lesion and need to handle in test time samples that may not contain a lesion but only background pixels. Our main contributions follow next.

Contributions and Paper Organization

Our main technical contributions are:

- We empirically study which combinations of CE and Dice losses work best in two lesion segmentation problems, for two different scenarios: a) when we assume that a lesion is always present on the image both in the training and in the test set, and b) when this assumption does not hold for the test set.
- We clarify ambiguous definitions in the literature about what is OoD data in binary segmentation and frame lesion segmentation on images that can potentially not contain foreground as an OoD generalization task.
- This paper also serves as a detailed description of our submission to the Diabetic Foot Ulcer Challenge 2022, held in conjunction with MICCAI 2022.

The rest of the paper is organized as follows: we first introduce notation to describe the CE and the soft dice losses in the context of binary segmentation, with different strategies to combine them into a single loss function. We then reflect on the definition of Out-of-Distribution data in binary image segmentation, and discuss the formulation of the no-foreground case as an OoD generalization task. This section is closed with the details of how we train segmentation models in this paper. We then move to the next section, where we provide a description of the data that we employ for our experiments. Following, we report comprehensive experimental results and discuss the performance of each loss function for the two considered segmentation problems. We finish the paper with some conclusions and general recommendations that derive from our analysis.

2   Methodology

In this section we first recall the formulation of the two loss functions used in this paper, and then introduce several simple loss combination mechanisms. Next, we formulate a definition of Out-of-Distribution data in the binary image segmentation context, and briefly describe other model training details.

2.1   The Cross-Entropy Loss, the Dice Loss, and Their Combinations

It is worthwhile to introduce first some notation. In the general case, let us consider:

- A neural network $\mathcal{U}_\theta$ taking image $\mathbf{x} = \{x_1, \ldots, x_{|\Omega|}\}$ defined on a spatial domain Ω, and generating a candidate segmentation $\mathcal{U}_\theta(\mathbf{x}) = \hat{\mathbf{y}} = \{\hat{y}_1, \ldots, \hat{y}_{|\Omega|}\}$.
- A pixel x_i belongs to one of C possible categories, $\hat{y}_i = (\hat{y}_i^1, \ldots, \hat{y}_i^C) \in [0,1]^C$, and we typically apply a softmax layer to the output of $\mathcal{U}_\theta$ so that $\sum_{c=1}^{C} \hat{y}_i^c = 1$, describing the probability of x_i being from of each class.
- In addition, we have a ground-truth image $\mathbf{y} = \{y_1, \ldots, y_{|\Omega|}\}$, where each pixel has been annotated with a single integer value, the label of the correct category: $y_i \in \{1, \ldots, C\}$.
- Finally, from the prediction $\hat{\mathbf{y}}$ we are eventually interested in generating a segmentation that contains also integer values, which we will denote by $\bar{\mathbf{y}}$. This is often achieved by applying the arg-max operation on each vector $\hat{y}_i$.

We can therefore perform a pixel-wise comparison between $\mathbf{y}$ and $\hat{\mathbf{y}}$ by means of a loss function, and iteratively update the parameters θ of our model so that the loss is minimized.

The Cross-Entropy Loss. A popular loss function for (pixel-wise) classification problems is the Log-loss, also known as *Cross-Entropy loss*, defined as:

$$\mathcal{L}_{CE}(\mathbf{y}, \hat{\mathbf{y}}) = -\frac{1}{|\Omega|} \sum_{i=1}^{|\Omega|} \sum_{c=1}^{C} \mathbb{1}^c(y_i) \cdot \log \hat{y}_i^c \tag{1}$$

where $\mathbb{1}^c(y_i)$ is zero unless pixel x_i belongs to class c, *i.e.* $y_i = c$, in which case $\mathbb{1}^c(y_i) = 1$. In other words, at a given pixel x_i we look at its log-probability of belonging to the correct class, and average over all pixels on the image.

For the particular case of binary segmentation problems - *foreground* versus *background*, which is the focus of this paper, there are two alternative approaches. We can just set $C = 2$ in the above description, so that for each pixel we have $\hat{y}_i = (\hat{y}_i^1, \hat{y}_i^2)$, but we can also realize that due to the constraint of probabilities summing up to one, $\hat{y}_i^2 = 1 - \hat{y}_i^1$. Therefore we could also keep only $\hat{y}_i^2$, or replace the softmax layer by a sigmoid operation that produces a single number in the $[0, 1]$ interval[1], and have the annotations be either 0 or 1, $y_i \in \{0, 1\}$. Simply speaking, the model now provides a single output $\hat{y}_i$ per pixel representing the probability of belonging to the foreground class. In this case, the *Binary Cross-Entropy* loss can be written as:

$$\mathcal{L}_{BCE}(\mathbf{y}, \hat{\mathbf{y}}) = -\frac{1}{|\Omega|} \sum_{i=1}^{|\Omega|} y_i \cdot \log(\hat{y}_i) + (1 - y_i) \cdot \log(1 - \hat{y}_i). \tag{2}$$

For foreground pixels ($y_i = 1$) only the predicted log-probability $\log(\hat{y}_i)$ of belonging to the foreground contributes to the loss, while for background pixels ($y_i = 0$) the contribution is the log-probability of belonging to the background, $\log(1 - \hat{y}_i)$. Therefore, Eqs. (1) and (2) are fully equivalent.

The Soft-Dice Loss. Another popular segmentation loss is the soft Dice, introduced for the binary case in [22] and later generalized to a (frequency-weighted) multi-class setting in [25]. It is a differentiable reformulation of the well-known Dice Similarity Score (DSC), a measure of set similarity given by:

$$\text{DSC}(X, Y) = \frac{2|X \cap Y|}{|X| + |Y|} \tag{3}$$

where the operator $|\cdot|$ represents a count of the number of elements in the set. If X and Y overlap perfectly, $\text{DSC}(X, Y) = 1$, and when the overlap decreases, then $|X \cap Y|$ diminishes while $|X| + |Y|$ is preserved, reducing the score.

In the binary segmentation case, we can use DSC to assess the performance of our model by measuring the similarity of the manual ground-truth $\mathbf{y}$ and a segmentation $\bar{\mathbf{y}}$. This can be done by defining $\mathbf{y} \cap \bar{\mathbf{y}} = \{i \in \Omega \mid y_i = \bar{y}_i = 1\}$ and $\mathbf{y} \cup \bar{\mathbf{y}} = \{i \in \Omega \mid y_i = 1 \text{ or } \bar{y}_i = 1\}$. In addition, we can write these quantities in terms of True/False Positives (TP, FP) and True/False Negatives (TN/FN) as $\hat{\mathbf{y}} \cap \mathbf{y} = \text{TP}$ and $\hat{\mathbf{y}} \cup \mathbf{y} = 2\text{TP} + \text{FP} + \text{FN}$, *i.e.*:

$$\text{DSC}(\mathbf{y}, \bar{\mathbf{y}}) = \frac{2|\mathbf{y} \cap \bar{\mathbf{y}}|}{|\mathbf{y}| + |\bar{\mathbf{y}}|} = \frac{2\text{TP}}{2\text{TP} + \text{FP} + \text{FN}}, \tag{4}$$

which shows that the Dice Similarity Score disregards True Negatives. This can be an advantage whenever TNs compose the majority of pixels on a segmentation, which is often the case in medical applications where large background regions are easily predicted by a model.

---

[1] In this case, the mechanism to build a segmentation $\bar{\mathbf{y}}$ by taking the argmax over $\hat{\mathbf{y}}$ would be equivalent to thresholding the sigmoid output with a value of $t = 0.5$.

In order to turn Eq. (4) into a loss function, we need it to take a binary ground-truth $\mathbf{y}$ and a continuous prediction $\hat{\mathbf{y}}$, and return a quantity that is differentiable, and decreases as the prediction improves [6]. This can be achieved by realizing that $|\mathbf{y} \cap \hat{\mathbf{y}}| = \sum_i y_i \cdot \hat{y}_i = \langle \mathbf{y}, \hat{\mathbf{y}} \rangle$, and $|\mathbf{y}| = \sum_i y_i \cdot y_i = \langle \mathbf{y}, \mathbf{y} \rangle$, so that we can write:

$$\mathcal{L}_{Dice}(\mathbf{y}, \hat{\mathbf{y}}) = 1 - \frac{2|\mathbf{y} \cap \hat{\mathbf{y}}|}{|\mathbf{y}| + |\hat{\mathbf{y}}|} = 1 - \frac{2\langle \mathbf{y}, \hat{\mathbf{y}} \rangle}{\langle \mathbf{y}, \mathbf{y} \rangle + \langle \hat{\mathbf{y}}, \hat{\mathbf{y}} \rangle}. \tag{5}$$

Finally, let us remark that, as opposed to the Cross-Entropy loss, the Dice loss is not defined at the pixel level but at the image level, and it degenerates if the ground-truth does not contain any foreground pixels, i.e. $\mathbf{y} = \mathbf{0}$. In this case, it will always be maximal regardless of the prediction $\hat{\mathbf{y}}$, but as the prediction improves ($\hat{\mathbf{y}} \to \mathbf{0}$), the denominator approaches zero, which can result in numerical instabilities during training.

Loss Combinations. The relationship between the CE and the soft Dice losses, and how to optimally combine them, has recently drawn attention in the research community both from a theoretical [19,28] and from a practical perspective [20]. It is common belief that the soft Dice loss can perform better in highly imbalanced scenarios and can result in DSC improvements [3], although it is known that this may come at the cost of poor calibration [21], which may require specific training techniques [11,15,18], post-processing methods, or fine-tuning [24].

In order to leverage the best of both worlds, one can also consider combining the two losses during training [20]. Indeed, even works that advocate for the use of the soft Dice loss over BCE, like [6] start the training with a preliminary stage in which models are trained until convergence with CE, and only then they are fine-tuned using soft Dice. In this work, we adopt five different loss combination strategies, namely:

1. **Only Binary Cross-Entropy.** We minimize $\mathcal{L}_{\text{BCE}}(\mathbf{y}, \hat{\mathbf{y}})$ alone.
2. **Loss addition.** A simple combination with equal weights on both losses:

$$\mathcal{L}_{\text{BCE+Dice}}(\mathbf{y}, \hat{\mathbf{y}})d = \mathcal{L}_{\text{BCE}}(\mathbf{y}, \hat{\mathbf{y}}) + \mathcal{L}_{\text{Dice}}(\mathbf{y}, \hat{\mathbf{y}}). \tag{6}$$

3. **Soft Fine-tuning.** We minimize a linear combination that starts giving full weight to BCE and ends up giving only weight to Dice, with intermediate weights linearly interpolated. At epoch $n = 0, 1, \ldots N$ this loss is given by:

$$\mathcal{L}_{\text{BCE}\leadsto\text{Dice}}(\mathbf{y}, \hat{\mathbf{y}}) = \frac{(N - n)}{N} \cdot \mathcal{L}_{\text{BCE}}(\mathbf{y}, \hat{\mathbf{y}}) + \frac{n}{N} \cdot \mathcal{L}_{\text{Dice}}(\mathbf{y}, \hat{\mathbf{y}}) \tag{7}$$

4. **Hard Fine-tuning.** We minimize the BCE loss in the first part of the training and switch to the Dice loss only in the last 10% of the training. For a model that trains over N epochs, at epoch n this is:

$$\mathcal{L}_{\text{BCE}\to\text{Dice}}(\mathbf{y}, \hat{\mathbf{y}}) = \begin{cases} \mathcal{L}_{BCE}(\mathbf{y}, \hat{\mathbf{y}}), & \text{if } n < 0.9 \cdot N \\ \mathcal{L}_{\text{Dice}}(\mathbf{y}, \hat{\mathbf{y}}), & \text{otherwise} \end{cases} \tag{8}$$

5. **Only soft Dice.** We minimize $\mathcal{L}_{\text{Dice}}(\mathbf{y}, \hat{\mathbf{y}})$ alone.

2.2 On the Definition of In-Distribution and Out-of-Distribution Data for Binary Segmentation Problems

The definition of what constitutes Out-of-Distribution (OoD) data is not clearly agreed in the machine learning community, let alone for binary segmentation tasks, so it is worth briefly clarifying the perspective we adopt here.

For classification problems, sometimes OoD data is defined as data that does not belong to any of the categories on which the model was trained [13], and the goal is to detect it and abstain from classifying it, a task known as **OoD Detection**. Other times, OoD data is considered as arising from some sort domain shift, *e.g.* different acquisition conditions, but not from a semantic shift - that is, OoD data belongs to the same set of categories as the in-distribution training data [26]. From this point of view, our model should be able to generalize to out-of-distribution data and classify correctly, a property known as **Robustness to OoD** in this context [12].

Image segmentation can be considered as a per-pixel classification problem, and in this case one can directly apply the above definition of OoD detection. The goal would be finding pixels on a test image that belong to anomalies not present in the training set [23]. A particularly popular approach to solve this problem is reconstruction-based models [2], in which an auto-encoder is tasked with rebuilding the input after reducing its dimensionality, and we expect it to fail in reconstructing OoD data. Since we can measure the reconstruction error pixel-wise, we are then able to localize OoD pixels individually. These could correspond to, for example, different pathologies like tumors or lesions, but also image quality degradations like localized visual artifacts [29].

On the other hand, in this paper we consider a more subtle OoD scenario that only exists for binary segmentation models, which are trained to receive an image $\mathbf{x}$ and produce a partition of its support Ω into foreground and background regions Ω^f and Ω^b. Specifically, we are interested in analyzing the performance of one such model on an in-distribution test set that contains the same kind of foreground objects, but also the degradation of model performance when test images *do not contain any foreground pixels*, that is, $\Omega = \Omega^b$.

2.3 Model Training Details

All the models in this paper were trained following our previous work on lesion segmentation [7], which we have applied successfully in several other biomedical segmentation challenges [1,14,27]. Specifically, our architectures are always a cascade of two encoder-decoder networks, the encoder being pretrained on ImageNet and the decoder being a Feature-Pyramid Network [17], and we consider several popular encoder architectures with increasing sizes. We optimize network weights by minimizing different loss combinations as described above, using an Adam optimizer with a learning rate of $l = 3e-4$ and a batch-size of 4. The learning rate is decayed to zero following a cosine law during training, which lasts for 40,000 optimization steps, which we empirically found enough for model

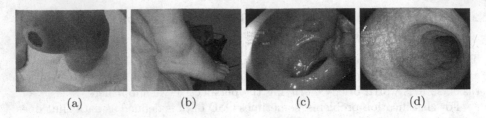

Fig. 1. The images in (a) and (c) contain a lesion marked in blue; (a) foot ulcer, (c) polyp. The other two images contain no foreground: (b) DFU remission case, (d) ulcerative colitis. We treat the latter as OoD data in this work. (Color figure online)

convergence. Images are resized to 640×512, which was the most common resolution in both of the considered datasets, and augmented during training with conventional image processing operations (random rotations, translations, scalings, vertical/horizontal flipping, contrast/saturation/brightness changes, etc.). We do not apply early stopping and simply keep the final model weights for testing purposes.

3   Experimental Results

We describe in this section our experimental validation on lesion segmentation in the presence of no-foreground OoD data.

3.1   Data and Performance Measures

Data in this paper comes from two different sources[2]: the Diabetic Foot Ulcer Segmentation Challenge DFUC2022 [16] and the Endotect 2021 challenge [14].

- The DFUC2022 dataset contains 4,000 640×480 images divided equally into a training and a test set [5,16]. The test set contains some cases of healthy individuals which are useful for testing our models in the absence of lesions. The organizers have also made available a small subset with 200 images, which contains no-foreground samples. Submission to the online grand-challenge website returns average performances on this test set, which we report below.
- The Endotect 2021 dataset contains 10,662 labeled endoscopic images capturing 23 different classes of findings, extracted from the Hyper-Kvasir dataset [4]. There are also segmentation masks for 1,000 images from the polyp class, which we use to train binary segmentation models. In test time, we use the polyp segmentation test set from the Endotect challenge, which contains 200 images containing polyps and corresponding binary masks, and we expand this set with another 20 images randomly sampled from the ulcerative collitis category in the Endotect dataset, *i.e.* images that do not contain a polyp but

---

[2] DFUC2022 Challenge: https://dfuc2022.grand-challenge.org/
Endotect 2020 Challenge: https://endotect.com/.

other kind of pathology. Figure 1 shows some examples of polyp and no-lesion cases, as well as examples from the DFUC2022 dataset.

In this paper we use a five-fold cross-validation approach: the training set is split in five subsets of equal size, one of them serving as a validation set. The validation set is rotated and training takes places five times, resulting in five sets of trained weights that we use for average performance estimation, and also for ensembling when generating predictions on the test set.

Performance is measured in terms of Dice Score (DSC) as in Eq. (4). Note that, following the DFUC2022 guidelines, we assign a score of 0 to predictions that contain a lesion when the ground-truth does not, and to predictions that contain no lesion when the ground-truth shows a lesion. A score of 1 is assigned to the case in which neither the ground-truth nor the prediction contain a lesion.

3.2 Performance Evaluation and Discussion

Performance on DFU Segmentation. We show in Table 1 results of 5-fold cross-validation experiments for DFU segmentation on **in-distribution data**, *i.e.* images that always contain a lesion. In this and the remaining cases of this section, we report average DSC for a range of CNN architectures, sorted top-to-bottom from smaller to larger in terms of learnable weights.

By looking at Table 1 we quickly realize that the training with $\mathcal{L}_{\text{BCE+Dice}}$ and $\mathcal{L}_{\text{BCE}\rightsquigarrow\text{Dice}}$ invariably leads to models achieving a stronger performance in terms of average DSC, regardless of the architecture. Indeed, any combination of the two base loss functions results in an improved performance when compared with using either $\mathcal{L}_{\text{BCE}}$ or $\mathcal{L}_{\text{Dice}}$ alone for driving the optimization. It is also worth noting that training with $\mathcal{L}_{\text{BCE}}$ leads to the worst results in almost all cases, unless for the largest encoders ResNet152 and ResNext101.

Results are strikingly different when it comes to testing on a mixture of in-distribution and **OoD data**, *i.e.* images that may not contain a lesion. Table 2

Table 1. Results with different architectures and loss functions for the task of **DFU segmentation**. For each model, <u>best</u> and **second best** are marked. Performance is DSC averaged over 5-fold training, only **in-distribution** data.

Enc./Loss	$\mathcal{L}_{\text{BCE}}$	$\mathcal{L}_{\text{Dice}}$	–	$\mathcal{L}_{\text{BCE}\rightarrow\text{Dice}}$	$\mathcal{L}_{\text{BCE}\rightsquigarrow\text{Dice}}$	$\mathcal{L}_{\text{BCE+Dice}}$
MobileNet	74.85 ± 0.48	77.45 ± 0.61	–	75.93 ± 0.88	$\mathbf{77.91 \pm 0.61}$	$\underline{\mathbf{77.94}} \pm 0.45$
ResNet18	74.73 ± 0.99	76.90 ± 0.67	–	76.20 ± 0.71	$\mathbf{77.57 \pm 0.65}$	$\underline{\mathbf{77.80}} \pm 0.6$
ResNet34	76.21 ± 0.56	77.73 ± 0.38	–	77.66 ± 0.49	$\mathbf{78.13 \pm 0.39}$	$\underline{\mathbf{78.69}} \pm 0.42$
ResNet50	76.37 ± 0.45	76.94 ± 0.49	–	77.04 ± 0.32	$\mathbf{78.02 \pm 0.51}$	$\underline{\mathbf{78.13}} \pm 0.27$
ResNeXt50	76.38 ± 1.05	77.72 ± 0.66	–	77.14 ± 0.97	$\mathbf{78.60 \pm 0.67}$	$\underline{\mathbf{78.82}} \pm 0.46$
ResNet101	76.55 ± 1.06	76.89 ± 0.70	–	77.36 ± 0.36	$\mathbf{78.04 \pm 0.59}$	$\underline{\mathbf{78.46}} \pm 0.67$
ResNeXt101	78.51 ± 0.75	77.32 ± 1.01	–	78.59 ± 0.52	$\underline{\mathbf{79.48}} \pm 0.43$	$\mathbf{78.97 \pm 0.40}$
ResNet152	77.07 ± 0.36	76.73 ± 0.52	–	77.95 ± 0.36	$\underline{\mathbf{78.52}} \pm 0.67$	$\mathbf{78.30 \pm 0.43}$

Table 2. Results with different architectures and loss functions for the task of **DFU segmentation**. For each model, **best** and **second best** are marked. Performance is DSC (% rejected images) achieved by a 5-fold ensemble on the hidden validation set, which included **OoD data**.

Enc./Loss	$\mathcal{L}_{BCE}$	$\mathcal{L}_{Dice}$	–	$\mathcal{L}_{BCE \to Dice}$	$\mathcal{L}_{BCE \leadsto Dice}$	$\mathcal{L}_{BCE+Dice}$
MobileNet	**65.73** (6.5%)	65.24 (2%)	–	65.35 (3.5%)	**<u>66.25</u>** (1.5%)	65.56 (2%)
ResNet18	66.05 (7%)	64.47 (1%)	–	**66.21** (5%)	66.10 (1%)	**<u>66.57</u>** (1%)
ResNet34	**<u>68.14</u>** (7%)	64.84 (1%)	–	**67.60** (4.5%)	66.53 (1.5%)	66.25 (1%)
ResNet50	**68.03** (7%)	65.21 (1%)	–	**<u>68.26</u>** (4.5%)	65.87 (1.5%)	64.99 (0.5%)
ResNeXt50	**67.65** (8%)	64.85 (0%)	–	**<u>67.99</u>** (4%)	66.51 (1.5%)	66.87 (1.5%)
ResNet101	67.92 (8%)	63.98 (1%)	–	**<u>68.29</u>** (5%)	65.09 (1%)	65.29 (2%)
ResNeXt101	**<u>68.70</u>** (5%)	64.45 (0.5%)	–	**68.03** (4%)	66.62 (1%)	65.49 (1%)
ResNet152	**<u>68.82</u>** (7%)	65.09 (0.5%)	–	**67.63** (4.5%)	66.21 (1%)	66.32 (1%)

shows the average DSCs for the same set of models after 5-fold ensembling and submission to the DFUC2022 Grand Challenge website, and we can see that the above trend is fully reverted. The weakest loss function, $\mathcal{L}_{BCE}$, becomes the strongest one in the presence of OoD images, largely outperforming the previously two best approaches, unless for the two smallest models. In addition, the Hard Fine-tuning approach using $\mathcal{L}_{BCE \to Dice}$, which was the loss combination with the lowest performance before, is now one of the most reliable solutions, achieving the best performance in three out of the nine considered models, and reaching the second position in another four cases. The poor performance of the $\mathcal{L}_{Dice}$ loss when optimized alone is also remarkable: irrespective of the size of the underlying architecture, the resulting DSC remains the worst of all considered approaches.

Performance on Polyp Segmentation. In the polyp segmentation problem we see a similar trend as for DFU segmentation. Namely, approaches that perform best for in-distribution data become again much worse when dealing with lesion-free images. As we can see in Table 3, training with $\mathcal{L}_{BCE \leadsto Dice}$ and $\mathcal{L}_{BCE+Dice}$ always results in the best performance, no matter the architecture. Perhaps the major difference with the previous case is the better average DSC achieved by $\mathcal{L}_{BCE}$, which now ranks as the third best option for most models. When we turn to testing on a combination of in-distribution and OoD data, we see again that both the Hard Fine-Tuning approach and the $\mathcal{L}_{BCE}$ loss, together with the $\mathcal{L}_{BCE \leadsto Dice}$ loss this time, achieve the highest average DSC, whereas the very popular $\mathcal{L}_{BCE+Dice}$ combination results in poor performance in most cases. Here again the optimization of $\mathcal{L}_{Dice}$ leads to the worst results across the board.

Submission to DFU Challenge In a last stage of the MICCAI DFUC2022 Challenge [?], the organizers requested submission on a large test set which included 2000 images, again with OoD data. We submitted the result of a five-

Table 3. Results with different architectures and loss functions for the task of **polyp segmentation**. For each model, <u>best</u> and **second best** are marked. Performance is DSC averaged over 5-fold training, only **in-distribution** data.

Enc./Loss	$\mathcal{L}_{BCE}$	$\mathcal{L}_{Dice}$	–	$\mathcal{L}_{BCE \to Dice}$	$\mathcal{L}_{BCE \rightsquigarrow Dice}$	$\mathcal{L}_{BCE+Dice}$
MobileNet	88.66 ± 0.84	88.63 ± 0.58	–	88.62 ± 0.76	$\underline{89.42} \pm 0.65$	$\mathbf{89.38} \pm 0.50$
ResNet18	89.25 ± 1.49	88.99 ± 1.07	–	89.28 ± 1.50	$\underline{89.96} \pm 1.00$	$\mathbf{89.55} \pm 1.12$
ResNet34	90.22 ± 0.51	89.02 ± 1.10	–	89.78 ± 0.89	$\mathbf{90.40} \pm 0.98$	$\underline{90.59} \pm 0.57$
ResNet50	90.36 ± 1.25	88.82 ± 0.70	–	90.13 ± 0.91	$\underline{90.71} \pm 0.90$	$\mathbf{90.67} \pm 0.90$
ResNeXt50	90.45 ± 0.93	89.47 ± 0.83	–	89.97 ± 0.73	$\mathbf{90.70} \pm 0.83$	$\underline{90.84} \pm 0.74$
ResNet101	90.49 ± 0.90	88.36 ± 0.62	–	90.30 ± 0.79	$\mathbf{90.88} \pm 0.87$	$\underline{90.96} \pm 0.75$
ResNeXt101	91.05 ± 1.13	88.33 ± 1.30	–	90.69 ± 1.13	$\mathbf{91.11} \pm 0.79$	$\underline{91.18} \pm 0.73$
ResNet152	90.47 ± 0.69	88.83 ± 1.35	–	$\mathbf{90.78} \pm 0.87$	$\underline{90.97} \pm 0.87$	90.64 ± 0.84

Table 4. Results with different architectures and loss functions for the task of **polyp segmentation**. For each model, <u>best</u> and **second best** are marked. Performance is DSC (% rejected images, in this case there were 9% OoD images) achieved by a 5-fold ensemble on the test set, which included **OoD data**.

Enc./Loss	$\mathcal{L}_{BCE}$	$\mathcal{L}_{Dice}$	–	$\mathcal{L}_{BCE \to Dice}$	$\mathcal{L}_{BCE \rightsquigarrow Dice}$	$\mathcal{L}_{BCE+Dice}$
MobileNet	86.64 (6%)	81.10 (1%)	–	**86.90** (6%)	<u>86.90</u> (6%)	83.91 (3%)
ResNet18	<u>87.72</u> (7%)	80.66 (0%)	–	**87.19** (6%)	86.65 (5%)	85.17 (4%)
ResNet34	**88.40** (7%)	82.89 (3%)	–	<u>89.78</u> (8%)	87.35 (5%)	85.78 (4%)
ResNet50	**88.40** (7%)	81.20 (1%)	–	<u>89.67</u> (8%)	88.30 (6%)	85.22 (4%)
ResNeXt50	88.85 (7%)	83.40 (2%)	–	<u>89.06</u> (7%)	**89.15** (6%)	85.22 (3%)
ResNet101	<u>89.33</u> (7%)	81.36 (1%)	–	**89.12** (8%)	88.15 (6%)	87.92 (6%)
ResNeXt101	**90.41** (8%)	81.54 (0%)	–	90.26 (8%)	<u>90.48</u> (8%)	89.65 (7%)
ResNet152	<u>90.48</u> (8%)	81.66 (1%)	–	89.77 (8%)	88.66 (7%)	88.67 (6%)

fold ensembling of our best model (ResNeXt-101) with Test-Time Augmentation. We also used a pseudo-labeling technique like the one described in [8] to improve generalization ability. Our final model reached the third place with DSC = 72.63, closely following the performance of the winning participant with a DSC = 72.87.

4   Conclusion

While it is well-known that adding $\mathcal{L}_{BCE}$ and $\mathcal{L}_{Dice}$ results in stronger in-distribution segmentation performance, the behavior of $\mathcal{L}_{BCE+Dice}$ on OoD data is unexplored. Our findings show that in that scenario it is not the best option due to poor robustness to the absence of lesions, and the Hard Fine-Tuning approach or even the standard $\mathcal{L}_{BCE}$ loss may be more reliable. Note however that the use of $\mathcal{L}_{BCE \to Dice}$ requires setting a hyper-parameter: the number of epochs needed in order to reach convergence. Finally, although it is known that $\mathcal{L}_{Dice}$

is poorly calibrated [21], it is also common knowledge that ensembling models results in improved calibration. For that reason, the performance of $\mathcal{L}_{\text{Dice}}$ in our OoD experiments is surprisingly disappointing, and we recommend avoiding the optimization of the soft Dice loss function alone in these cases (Table 4).

Acknowledgments. This work was supported by a Marie Skłodowska-Curie Fellowship (No. 892297) and by Australian Research Council grants (DP180103232 and FT190100525).

References

1. Ali, S., et al.: Assessing generalisability of deep learning-based polyp detection and segmentation methods through a computer vision challenge (2022). arXiv:2202.12031 [cs]
2. Baur, C., et al.: Autoencoders for unsupervised anomaly segmentation in brain MR images: A comparative study. Med. Image Anal. **69**, 101952 (2021). https://doi.org/10.1016/j.media.2020.101952
3. Bertels, J., et al.: Optimizing the dice score and Jaccard index for medical image segmentation: theory and practice. In: Shen, D., et al. (eds.) MICCAI 2019. LNCS, vol. 11765, pp. 92–100. Springer, Cham (2019). https://doi.org/10.1007/978-3-030-32245-8_11
4. Borgli, H., et al.: HyperKvasir, a comprehensive multi-class image and video dataset for gastrointestinal endoscopy. Sci. Data **7**(1), 283 (2020). https://doi.org/10.1038/s41597-020-00622-y
5. Cassidy, B., et al.: The DFUC 2020 dataset: analysis towards diabetic foot ulcer detection. touchREVIEWS Endocrinol. **17**, 5–11 (2021). https://doi.org/10.17925/EE.2021.17.1.5, https://www.touchendocrinology.com/diabetes/journal-articles/the-dfuc-2020-dataset-analysis-towards-diabetic-foot-ulcer-detection/1
6. Eelbode, T., et al.: Optimization for medical image segmentation: theory and practice when evaluating with dice score or Jaccard index. IEEE Trans. Med. Imaging **39**(11), 3679–3690 (2020). https://doi.org/10.1109/TMI.2020.3002417
7. Galdran, A., et al.: Double encoder-decoder networks for gastrointestinal polyp segmentation. In: ICPR International Workshops and Challenges, pp. 293–307 (2021). https://doi.org/10.1007/978-3-030-68763-2_22
8. Galdran, A., et al.: State-of-the-art retinal vessel segmentation with minimalistic models. Sci. Rep. **12**(1), 6174 (2022). https://doi.org/10.1038/s41598-022-09675-y
9. Galdran, A., et al.: Test time transform prediction for open set histopathological image recognition (2022). arXiv:2206.10033 [cs]
10. Goyal, M., Yap, M.H., Reeves, N.D., Rajbhandari, S., Spragg, J.: Fully convolutional networks for diabetic foot ulcer segmentation. In: 2017 IEEE International Conference on Systems, Man, and Cybernetics (SMC), pp. 618–623. IEEE (2017)
11. Gros, C., et al.: SoftSeg: advantages of soft versus binary training for image segmentation. Med. Image Anal. **71**, 102038 (2021). https://doi.org/10.1016/j.media.2021.102038
12. Hendrycks, D., et al.: The many faces of robustness: a critical analysis of out-of-distribution generalization. In: ICCV (2021)
13. Hendrycks, D., Gimpel, K.: A baseline for detecting misclassified and out-of-distribution examples in neural networks. In: ICLR (2017)

14. Hicks, S.A., Jha, D., Thambawita, V., Halvorsen, P., Hammer, H.L., Riegler, M.A.: The EndoTect 2020 challenge: evaluation and comparison of classification, segmentation and inference time for endoscopy. In: Del Bimbo, A., et al. (eds.) ICPR 2021. LNCS, vol. 12668, pp. 263–274. Springer, Cham (2021). https://doi.org/10.1007/978-3-030-68793-9_18

15. Islam, M., Glocker, B.: Spatially varying label smoothing: capturing uncertainty from expert annotations. In: Feragen, A., Sommer, S., Schnabel, J., Nielsen, M. (eds.) IPMI 2021. LNCS, vol. 12729, pp. 677–688. Springer, Cham (2021). https://doi.org/10.1007/978-3-030-78191-0_52

16. Kendrick, C., et al.: Translating clinical delineation of diabetic foot ulcers into machine interpretable segmentation (2022). arXiv:2204.11618 [cs, eess]

17. Lin, T.Y., et al.: Feature pyramid networks for object detection. In: Conference on Computer Vision and Pattern Recognition (CVPR) (2017)

18. Liu, B., et al.: The devil is in the margin: margin-based label smoothing for network calibration. In: CVPR, pp. 80–88 (2022)

19. Liu, B., et al.: The hidden label-marginal biases of segmentation losses (2022). arXiv:2104.08717 [cs]

20. Ma, J., et al.: Loss odyssey in medical image segmentation. Med. Image Anal. **71**, 102035 (2021). https://doi.org/10.1016/j.media.2021.102035

21. Mehrtash, A., et al.: Confidence calibration and predictive uncertainty estimation for deep medical image segmentation. IEEE Trans. Med. Imaging **39**(12), 3868–3878 (2020). https://doi.org/10.1109/TMI.2020.3006437

22. Milletari, F., et al.: V-net: fully convolutional neural networks for volumetric medical image segmentation. In: 2016 Fourth International Conference on 3D Vision (3DV) (2016). https://doi.org/10.1109/3DV.2016.79

23. Popescu, S.G., Sharp, D.J., Cole, J.H., Kamnitsas, K., Glocker, B.: Distributional Gaussian process layers for outlier detection in image segmentation. In: Feragen, A., Sommer, S., Schnabel, J., Nielsen, M. (eds.) IPMI 2021. LNCS, vol. 12729, pp. 415–427. Springer, Cham (2021). https://doi.org/10.1007/978-3-030-78191-0_32

24. Rousseau, A.J., et al.: Post training uncertainty calibration of deep networks for medical image segmentation. In: IEEE ISBI, pp. 1052–1056 (2021). https://doi.org/10.1109/ISBI48211.2021.9434131

25. Sudre, C.H., Li, W., Vercauteren, T., Ourselin, S., Jorge Cardoso, M.: Generalised dice overlap as a deep learning loss function for highly unbalanced segmentations. In: Cardoso, M.J., et al. (eds.) DLMIA/ML-CDS -2017. LNCS, vol. 10553, pp. 240–248. Springer, Cham (2017). https://doi.org/10.1007/978-3-319-67558-9_28

26. Tran, D., et al.: Plex: towards reliability using pretrained large model extensions (2022). https://doi.org/10.48550/arXiv.2207.07411

27. Wang, C., et al.: FUSeg: the foot ulcer segmentation challenge (2022). arXiv:2201.00414 [cs, eess]

28. Yeung, M., et al.: Unified focal loss: generalising dice and cross entropy-based losses to handle class imbalanced medical image segmentation. Comput. Med. Imaging Graph. **95**, 102026 (2022). https://doi.org/10.1016/j.compmedimag.2021.102026

29. Zimmerer, D., et al.: MOOD 2020: a public benchmark for out-of-distribution detection and localization on medical images. IEEE Trans. Med. Imaging 1 (2022). https://doi.org/10.1109/TMI.2022.3170077

Capture the Devil in the Details via Partition-then-Ensemble on Higher Resolution Images

Yung-Han Chen[1]([✉]) [iD], Yi-Jen Ju[2], and Juinn-Dar Huang[1] [iD]

[1] Institute of Electronics, National Yang Ming Chiao Tung University,
Hsinchu, Taiwan
james0710121.ee07@nycu.edu.tw
[2] Computer Science, National Tsing Hua University,
Hsinchu, Taiwan

Abstract. For the Diabetic Foot Ulcer Challenge 2022 (DFUC2022) hosted by MICCAI 2022, we built a machine learning model based on the architecture of TransFuse [20] to accomplish the segmentation task. The TransFuse model combines Transformers and convolutional neural networks (CNNs), taking advantage of both local and global features. In this paper, we propose a modification to the data flow in encoder necks for decoding features in the higher resolution level, and in fusion modules for more efficient attention. Furthermore, to minimize the information loss as a result of resizing, we propose new techniques in both training and testing algorithms. Firstly, a region proposal network (RPN) is introduced from object detection methods and is used at the image pre-processing phase. It crops fixed size images from origin images, so that the high resolution input can be fed into TransFuse. We also applied test-time augmentation following a similar concept to RPN. We crop fixed size images at each corner and use edge pooling to ensemble them properly.

Keywords: Transformer · CNN · Medical image segmentation · Object detection · Ensemble

1 Introduction

Diabetic foot ulcer occurs in many diabetic patients. Without close monitoring and management, it can lead to infection or even amputation. Machine learning-based computer vision tools can help patients monitor changes in their ulcers and manage their wounds accordingly. In particular, image segmentation of ulcer photos can help patients quickly and precisely follow changes in shape and size of their ulcers. MICCAI 2022 offers the largest segmentation dataset, namely DFUC2022 [12], for training of such tools. In this paper, we propose a method to train on DFUC2022 and similar datasets in order to achieve high performance in segmenting diabetic foot ulcers.

M. H. Yap et al. (Eds.): DFUC 2022, LNCS 13797, pp. 52–64, 2023.
https://doi.org/10.1007/978-3-031-26354-5_5

Our method is divided into two stages, the region-proposal stage and the segmentation stage. The region-proposal stage takes inspiration from region-proposal networks often used for object detection tasks [6]. We train a model that learns to identify regions that may contain the ulcer. The segmentation model takes as input the regions identified by the region-proposal model and outputs the final segmentation masks. Such two-stage design minimizes the need to resize the original image into that compatible with the segmentation models. The region-proposal model crops a region that likely contains the ulcer to the exact dimensions on which the segmentation model was trained, so that the cropped image can be input into the subsequent segmentation model without resizing or any other preprocessing. The benefits of our two-stage method are two-fold. Firstly, the resolution of the original image is preserved; every pixel can be used as information in the segmentation stage. Secondly, noisy background is removed before the image is used for segmentation. This can potentially lead to more parameters in the segmentation model being used for detailed, high-resolution segmentation of the ulcer, instead of directing some of the parameters to distinguishing irrelevant background.

Since the RPN is a learned model and is, although seldom, subject to error, we needed additional cropped images to ensure that images input into the segmentation model do indeed contain the ulcer. A total of six cropped images per original image are used for training. Four cropped images are obtained from cropping a 384×384 region on each of the four corners, the union of which covers the entire area of the original image. One image is the output of the RPN. And one final image is the original image resized to 384×384, which allows the model to obtain spatial information from the whole image. To properly ensemble these six images, we devised an edge pooling technique that applies a different weight to the edge detection results from each of the six images, according to the certainty of each result. Specifically, edge detection from high-resolution images such as the RPN output or those cropped from four corners of the original image should be given higher weights. In contrast, low-resolution edge detection results from the resized original image should be given lower weights.

For the segmentation stage, we built our model in light of the outstanding performance of TransFuse in several medical segmentation tasks [20]. Combining CNNs and Transformers, it has a wider receptive field than many previous works on medical segmentation. However, since DFUC2022 is a small-region segmentation task, some modifications to TransFuse are needed to improve its performance in edge detection. Furthermore, we also added additional upsampling modules in the encoder instead of directly upsampling at the end of decoder, so that the feature maps from the encoder can be fed into the decoder with more spatial information.

2   Proposed Method

Our proposed segmentation network architecture is shown in Fig. 2. We preserve the main structure of the TransFuse model with several modifications.

Firstly, we maintained the two-branch structure in the encoder but replaced each branch with a different backbone. Secondly, we designed a new fusion module for the new backbones to achieve more efficient combination of features from the two branches. Finally, the RPN is introduced and thereby a two-stage training method, as shown in Fig. 1. The training and testing algorithms for the segmentation stage are optimized to take advantage of, or make up for, the result from the first object detection stage, including corner cropping, loss function refinement in the training phase, and ensemble weighting in the testing phase.

2.1 Two-Stage Architecture

Our framework consists of two stages in sequence. The former is the region-proposal stage. It crops the original image and feeds the image to the later segmentation stage with minimum resizing. The segmentation stage then inferences the image.

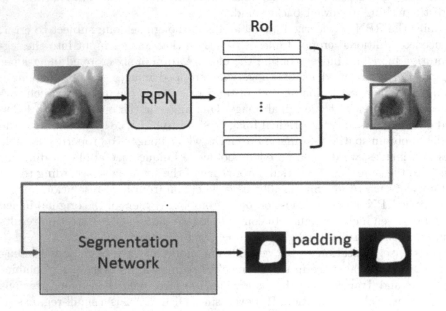

Fig. 1. Overview of our two-stage architecture. The output of the RPN is a list of bounding boxes. We obtain a super-bounding box from the list of bounding boxes, and crop a fixed-size image as input for the segmentation stage. The segmentation output will be padded to the original image size.

RPN. Resizing images is often performed when training or fine-tuning on pre-trained networks, because the size of the images in the dataset at hand is not necessarily the same as that on which the pre-trained networks were trained. However, resizing images compromises the image resolution and can lead to loss in information. The region-proposal network, or RPN [6], is introduced because 89% of ulcers are less than 5% of the total image size in the DFUC2022 dataset [12]. It seems unnecessary, therefore, to resize the whole image into a size compatible with the segmentation network while only a portion of the image is relevant to segmentation. As shown in Fig. 1, the RPN is a separate network that precedes the segmentation network in training. It identifies potential regions that may contain the ulcer, and crops the image to retain only such region, then inputs the cropped image to the segmentation network. The rest of this section will describe the detailed implementation of the RPN. We trained the RPN using binary masks provided in DFUC2022 transformed into bounding boxes as ground truth. The trained RPN outputs a list of bounding boxes that may contain the ulcer, but because the segmentation network only takes one image as input, we had to devise a way to retain only one bounding box. We propose to extract a single 384 × 384 bounding from a list of potential ulcer regions by a method of "super-bounding box". We first take a single super-bounding box specified by the outermost four corners of the collection of bounding boxes. Namely, the upper left and upper right coordinates of the super-bounding box take on values of the minimum of y-coordinates of all bounding boxes, and the minimum and maximum x-coordinates of all bounding boxes, respectively. Similarly, the bottom left and bottom right coordinates of the super-bounding box take on values of the maximum y-coordinates of all bounding boxes, and minimum and maximum x-coordinates of all bounding boxes, respectively. The super-bounding box can vary greatly in size. However, in order to use the image cropped against the super-bounding box to fine-tune a pre-trained segmentation network, we need an image of size 384 × 384. There are two cases to consider: where the resultant image is smaller than 384 × 384 and where it is larger than 384 × 384. To avoid resizing, when the resultant image is smaller than 384 × 384, we take a region that is 192 pixels away in all four directions from the center of the super-bounding box to obtain the desired bounding box of size 384 × 384. On the other hand, if the super-bounding box is larger than 384 × 384, we resize it to 384 × 384. If any such region is out of bound of the original image, we simply translate it to a position where at least one of the coordinates is on the edge of the original image. Such workflow minimizes the need to resize the original image; resizing is only required for a super-bounding larger than 384 × 384, which occurs only rarely. As a result, the high resolution of the original image is preserved as much as possible, while pre-trained segmentation networks can still be used to improve training efficiency.

Transformer Branch. Transformers have entered the realm of computer vision since Vision Transformer [4], and have been optimized by Deit [16]. However, the depth of the network is limited by computational resources. Following the publication of Swin-Transformer [14], the pyramid and window-based architectures led to increase in network depth and, in turn, accuracy in image classification. We selected Cswin-Transformer as the backbone of the Transformer branch [3] because it is the state-of-the-art (SOTA) model with publicly available pre-

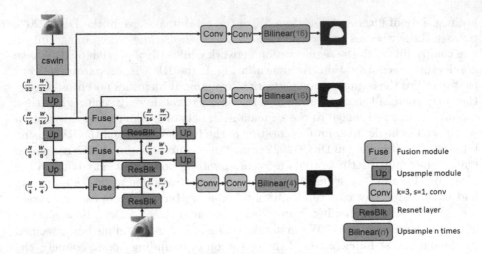

Fig. 2. Overview of our modified TransFuse model. The Transformer-branch encoder is CSwin-Base and the CNN-branch is ResNet-50. They will be fused by Fusion modules in 3 different resolutions. Because of the additional upsampling module applied on the Transformer branch, feature maps can be decoded in higher resolution than that in the original TransFuse model.

trained weights. Training without pre-trained weights causes over 2% drop in mean Dice (mDice).

CNN Branch. For the CNN branch, ResNet-50 and HarDNet-68 [2,10] are selected as candidates for the backbone. The main goal of the CNN branch is to extract local features, so deepness is not necessary. Furthermore, limiting the model size is helpful to decrease the memory overhead during training. Through our experiments, we found that ResNet-50 has higher performance in validation, so it was adopted as the CNN-branch backbone.

Fusion Module. The function of the Fusion module is to apply attention to and combine the outputs of the Transformer and CNN branches. Two kinds of attention modules are applied: the squeeze-and-excite (SE) block, a channel-attention technique, and the convolutional block attention module (CBAM), a spatial-attention technique [11,19]. Because the feature maps from CNN branches contain local information but lack global features, the CBAM block is more suitable. In contrast, since maps from the Transformer branch contain global features but less local features, the SE block is a better fit. This is in contrast to the design of the fusion module in the original TransFuse model, where the SE block is used for the CNN branch and the CBAM block is used for the Transformer branch. In our experiments, the CNN-CBAM and Transformer-SE architecture showed more than 2% improvement in mDice from that of the original Transfuse model. The module connections are shown in Fig. 3.

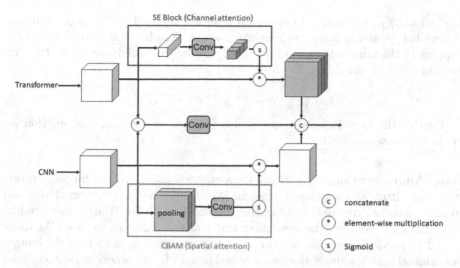

Fig. 3. Overview of our Fusion module. An SE block is applied to the Transformer branch for channel attention and CBAM is applied to the CNN branch for spatial attention. There is also a residual link to preserve part of the input.

2.2   Training Algorithm

DFUC2022 has only 2000 images for training [12], so training should be optimized to avoid overfitting. In this paper, we present some modifications on the loss function and methods of data augmentation.

$$w = 1 + 5\times \mid AvgPool2d(G) - G \mid \tag{1}$$

$$w_{scale} = 1 + 0.2 \times (G > 0.5) \tag{2}$$

$$w_{total} = w \times w_{scale} \tag{3}$$

Loss Function. We took inspiration from the loss function in PraNet, $L = L_{BCE}^w + L_{IoU}^w$ [5], where w indicates the weights for each pixel, and the weights on the edges are larger according to the definition in Eq. 1, P is the binary prediction map and G is the binary ground truth map. According to the positive area percentage occupied in DFUC2022 [12], we decrease the kernel size of average pooling from 31 to 7, so that the edges will be sharper. Furthermore, there is an additional weight w_{scale} used to improve the false negative rate. The imbalance between positive and negative areas can be smoothed with Eq. 2, and our final weight map of the loss function is shown in Eq. 3.

$$L_{IoU}^w = \frac{P^w}{P^w + G^w - P^w \times G^w} \tag{4}$$

$$L_{IoU}^w = \frac{P^w}{\mid P^w - G^w \mid + G^w} \tag{5}$$

As for L_{IoU}^w, the original expression is shown in Eq. 4. P^w and G^w indicate the prediction and ground truth weighted by w. We observed that in Eq. 4, P^w appears in the numerator as well as the denominator, which can make training unstable. We modified it to Eq. 5.

$$L = (L_{BCE}^{w_{total}} + L_{IoU}^{w_{total}}) \times \mid P - G \mid^2 \qquad (6)$$

Finally, the focal loss technique is also applied, so the final loss function we applied is shown as Eq. 6.

Data Augmentation. Traditional data augmentation such as flipping, rotating, color jitter, etc. requires images in DFUC2022 be resized from 480×640 to 384×384 [12], due to the dimension specification of the Transformer branch. This, however, reduces the resolution and leads to information loss. As mentioned in previous sections, we trained our segmentation model on six images per original image, namely the four cropped images from corners, an output from the RPN, and the original image resized. The five additional images serve as data augmentation, with the additional benefit of preserving the high resolution of the original image.

Multi-scaling. Medical images often have large variance in the positive area and image resolution. Therefore, we introduce the multi-scale training, published by PraNet [5], into our training scheme. The sizes of model inputs are resized to 1.25, 1 and 0.75 times of their original sizes in each epoch of training. However, unlike previous implementations of this method, our input size is limited by the Transformer branch and only 384×384 images can be fed into our model. As a result, we crop the images around the center if they have been upsampled and zero-pad around the edge if they have been downsampled, thereby obtaining fixed-size images.

2.3   Testing Algorithm

We have verified that optimization in the training phase improves the robustness of a single model, and more kinds of testing methods can be implemented to improve results. To take full advantage of data augmentation from corner cropping, we ensemble all cropped images with an edge-weighting method. Different from cropped images, the output of the resized images have to be upsampled from 384×384 to 480×640, which indicates its edges are distorted again. The edge-weighting method helps to vanish weights of the resized images on edges to make high resolution outputs dominate the results.

Cropping Ensemble and Edge Weighting. The testing results usually have a high false positive rate, especially on the edges of the ulcers, because the training dataset lacks negative samples in which no foot ulcer occurs. Now there are four more high-resolution inputs from corner-cropped images, and they divide the original image into 9 regions. After summing up four outputs, we get prediction map $res1$. A weight map $w1$, which is the same dimension as the original image, indicates the number of cropped images by which each pixel is included. The output from the resized original image is named $res0$, and is used to obtain weight $w0$ with Eq. 7.

$$w0 = 2\times \mid AvgPool2d(Sigmoid(res0)) - 0.5 \mid \tag{7}$$

According to Eq. 7, the weights of edges predicted from $res0$ will be eliminated, and the results will be dominated by high resolution images. The final binary prediction map res is obtained from Eq. 8.

$$res = Sigmoid(\frac{res1 + w0 \times res0}{w1 + w0}) > threshold \tag{8}$$

Test-Time Augmentation. We apply test-time augmentation through different kinds of data augmentation methods during testing [18], such as horizontal flipping, vertical flipping, clock-wise rotation by $90°$, and counter-clockwise rotation by $90°$. Each method of augmentation is accompanied by the cropping ensemble method mentioned above, thereby producing five different results. Finally, we obtain the final prediction via a voting among these five results.

3   Training Details

Prior to training the segmentation network, we trained a RPN with a ResNet-50 backbone for 10,000 iteration [15]. The AdamW optimization algorithm was employed to optimize parameters with a learning rate of 10^{-5}. The batch size of each iteration is 32, sampled randomly from 1,800 images in the training dataset, and the rest of 200 images were used as validation data.

As for segmentation, there are two stages to train. We firstly loaded ImageNet pre-trained weights of the two backbones [3,10], and trained our models with 1,800 images using the AdamW algorithm, setting the learning rate to 3×10^{-5}, and batch size to 8 for 100 epochs. The loss function is the same as the one defined in the PraNet [5]. The set of parameters performing the best during validation was saved to be the pre-trained weights for the next phase.

With the best-performing parameters from the first stage of training, we trained with the total 2,000 images for 50 epochs, or 20 epochs if the multi-scaling method is applied. In this stage, the batch size is set to 12, in which 4 of them are the resized input, another 4 of them are the cropped images from one of corners, and the rest are cropped according to the RPN results. We also used two different optimizers. One is the SGD optimizer with a momentum of 0.9 and a learning rate of 10^{-4}, scheduled by a cosine-annealing scheduler. It is only employed for the CNN-branch backbone in our TransFuse model. The other one is the AdamW optimizer and its learning rate is set to 3×10^{-5}. It is scheduled by another cosine annealing scheduler and employed for the rest of the parameters in the model. The learning rates of both optimizers would decay to 0.1 of the initial learning rate at the end of training.

4   Experiments and Results

Table 1. Experiment results in validation with different encoder backbones and additional upsample module at the end of the Transformer branch.

Transformer	CNN	Upsample	mIoU
Swin-B	HarDNet-68	No	0.6958
Swin-B	ResNet-50	No	0.6991
CSwin-B	HarDNet-68	No	0.6991
CSwin-B	ResNet-50	No	**0.7043**
CSwin-B	HarDNet-68	Yes	0.7023
CSwin-B	ResNet-50	Yes	**0.7090**

In the early stages of the dataset development, the number of images shared by the organisers is relatively small [7–9]. Since the establishment of challenges [1, 12,17], the number of images shared by the organisers increased. For DFUC2022 [12], we split the 2,000 images in the training set randomly into 1,800 training images and 200 validation images. The performance against the validation set was used to determine our model architecture. We based our selection of CNN and Transformer backbones in the encoder solely on mean IoU (mIoU), and test the reliability of the modified TransFuse architecture shown in Fig. 2. The experiment results are shown in Table 1.

We later modified the training settings and considered both the mIoU and mDice metrics, while still saving the best parameters tested in the 200 validation images. Firstly, we trained from scratch with the model selected based on performance metrics recorded in Table 1, and compared the experiment results of this model with different learning rates and optimizers. Secondly, we loaded the best weights obtained while training from scratch and trained on these weights with corner cropping augmentation. The experiment results are shown in Table 2.

In the final stages of the challenge, optimizations include not only testing but training algorithms, and mDice shown in the validation stage on the leaderboard became the primary metric. The experiment results are shown in Table 3.

Firstly, in the testing phase, we tried averaging before sigmoid or sigmoid before averaging for ensemble of the cropped images, and we found that the former performed better because the nonlinearity of sigmoid helps high-confidence results to dominate the less certain predictions. Secondly, we got pooling weights from prediction results of the resized input according to Eq. 7, and tested the performance. Finally, test-time augmentation is applied with counterclockwise rotating, clockwise rotating, horizontal flipping, and vertical flipping.

Table 2. Experiment results in validation with different training settings. The experiments in the upper table are trained from scratch, and the setting in the 2nd row performs the best on the validation dataset, so it was selected as the pre-trained weight for the experiments in the lower table. Both rows in the lower table are trained with corner cropping augmentation.

Resnet lr	Resnet optim	Other lr	Other optim	Post-train	mIoU	mDice
3×10^{-5}	AdamW	3×10^{-5}	AdamW	No	0.7090	0.8016
10^{-4}	AdamW	3×10^{-5}	AdamW	No	**0.7084**	**0.8034**
10^{-4}	SGD	3×10^{-5}	AdamW	No	0.6961	0.7962
10^{-4}	AdamW	3×10^{-5}	AdamW	Yes	0.7140	0.8085
10^{-4}	SGD	3×10^{-5}	AdamW	Yes	**0.7307**	**0.8190**

Table 3. Experiment results in the validation stage shown in the top table are with different testing methods, including ensemble method, and pooling weight. The middle section shows the improvements when using the fine-tuned loss function and test-time augmentation (TTA). The last section shows the results with additional weight w_{scale} and multi-scaling training method.

Ensemble	Pooling weight	Fine-tuned loss	TTA	w_{scale}	Multi-scaling	mDice
sigmoid first	No	No	No	No	No	0.6896
sigmoid first	No	No	No	No	No	0.6902
average first	Yes	No	No	No	No	0.6919
average first	Yes	No	No	No	No	**0.6927**
average first	Yes	Yes	No	No	No	0.6952
average first	Yes	Yes	Yes	No	No	**0.7023**
average first	Yes	Yes	Yes	Yes	No	**0.7097**
average first	Yes	Yes	Yes	No	Yes	0.7034

As for the training phase, we fine-tuned the kernel size of $AvgPool2d$ in Eq. 1 from 31 to 7 and applied our L_{IoU}^{w} and focal loss as shown in Eq. 5 and Eq. 6.

Table 4. Different from the best combination of training methods we obtained from the Valid Stage, the Testing Phase performed the best without additional training methods. Furthermore, the threshold of TTA should not be 1 because the false positive issue will dominate the performance.

Threshold	w_{scale}	Multi-scaling	mDice	FPE	FNE
2	No	No	**0.7269**	0.2045	**0.2388**
1	No	No	0.7254	**0.1847**	0.2582
2	Yes	No	0.7253	0.2053	0.2395
1	Yes	No	0.7226	0.1886	0.2581
2	No	Yes	0.7243	0.2052	0.2398
1	No	Yes	0.7235	0.1851	0.2568

The additional w_{scale} shown in Eq. 2 and multi-scaling method in the training phase are tested following by the fine-tuned loss function.

Table 5. Extension of Table 4. We modified the loss function to the original one, and repeated the combination of training methods. The performance improved further than the Testing Phase, so we concluded that the loss function is irregularly relative to the dataset.

Focal loss	w_{scale}	Multi-scaling	mDice	FPE	FNE
No	No	No	0.7270	0.2152	0.2290
Yes	No	No	0.7280	**0.2136**	0.2290
Yes	FP	No	0.7261	0.2214	**0.2257**
Yes	FN	No	0.7247	0.2156	0.2333
Yes	No	Yes	**0.7285**	0.2151	0.2268

According to the final validation results, the models shown in the last three rows in Table 3 were selected as candidates to test all 2,000 test set images in the testing stage. The voting threshold of the TTA method in the testing phase were set to 1 or 2. Therefore, there are 6 results in total in the testing stage shown in the Table 4. In contrast to the conclusion in Table 3, the best results were obtained without the additional w_{scale} or multi-scaling method in the training phase. This might have been due to the difference between datasets, so the training algorithm should be fine-tuned further based on this result (Table 5).

5   Conclusion

The methods applied in this paper are mostly inspired by the concept of object detection, where we focus our attention on portions of the image more likely to be relevant to segmentation [6], rather than process the whole image at once. Still, the global view is necessary. We therefore devised methods of image ensemble, combining results from both local and global views of the image. We did not, however, apply model ensemble because it is too impractical and causes too much memory overhead. With several training optimizations, including data augmentation with cropping and the RPN results, fine-tuning optimizers, and the focal loss technique with self-defined IoU loss function, the ensemble method works successfully in the testing phase. Furthermore, the additional upsampling modules in our modified TransFuse model contribute significantly since the feature maps are decoded in higher resolution [20], preserving more spatial information. Finally, the pooling weight method and test-time augmentation in the testing phase are helpful to increase overall confidence of prediction and lead to better performance.

Acknowledgments. Supported by TWCC, and MOST.

References

1. Cassidy, B., et al.: DFUC 2020: Analysis Towards Diabetic Foot Ulcers Detection (2020). arXiv:2004.11853
2. Chao, P., Kao, C.Y., Ruan, Y.S., Huang, C.H., Lin, Y.L.: Hardnet: a low memory traffic network. In: Proceedings of the IEEE/CVF International Conference on Computer Vision, pp. 3552–3561 (2019)
3. Dong, X., et al.: Cswin transformer: a general vision transformer backbone with cross-shaped windows. In: Proceedings of the IEEE/CVF Conference on Computer Vision and Pattern Recognition, pp. 12124–12134 (2022)
4. Dosovitskiy, A., et al.: An image is worth 16x16 words: Transformers for image recognition at scale (2020). arXiv preprint arXiv:2010.11929
5. Fan, D.-P., et al.: PraNet: parallel reverse attention network for polyp segmentation. In: Martel, A.L., et al. (eds.) MICCAI 2020. LNCS, vol. 12266, pp. 263–273. Springer, Cham (2020). https://doi.org/10.1007/978-3-030-59725-2_26
6. Girshick, R.: Fast r-cnn. In: Proceedings of the IEEE International Conference on Computer Vision, pp. 1440–1448 (2015)
7. Goyal, M., Reeves, N.D., Davison, A.K., Rajbhandari, S., Spragg, J., Yap, M.H.: DFUNet: convolutional neural networks for diabetic foot ulcer classification. IEEE Trans. Emerg. Topics Comput. Intell. 4(5), 728–739 (2018)
8. Goyal, M., Reeves, N.D., Rajbhandari, S., Yap, M.H.: Robust methods for real-time diabetic foot ulcer detection and localization on mobile devices. IEEE J. Biomed. Health Inf. 23(4), 1730–1741 (2018)
9. Goyal, M., Reeves, N., Rajbhandari, S., Ahmad, N., Wang, C., Yap, M.H.: Recognition of Ischaemia and infection in diabetic foot ulcers: dataset and techniques. Comput. Biol. Med. 117, 103616 (2019)
10. He, K., Zhang, X., Ren, S., Sun, J.: Deep residual learning for image recognition. In: Proceedings of the IEEE Conference on Computer Vision and Pattern Recognition, pp. 770–778 (2016)
11. Hu, J., Shen, L., Sun, G.: Squeeze-and-excitation networks. In: Proceedings of the IEEE Conference on Computer Vision and Pattern Recognition, pp. 7132–7141 (2018)
12. Kendrick, C., et al.: Translating Clinical Delineation of Diabetic Foot Ulcers into Machine Interpretable Segmentation (2022). arXiv:2204.11618
13. Lin, T.Y., Goyal, P., Girshick, R., He, K., Dollár, P.: Focal loss for dense object detection. In: Proceedings of the IEEE International Conference on Computer Vision, pp. 2980–2988 (2017)
14. Liu, Z., et al.: Swin transformer: hierarchical vision transformer using shifted windows. In: Proceedings of the IEEE/CVF International Conference on Computer Vision, pp. 10012–10022 (2021)
15. Ren, S., He, K., Girshick, R., Sun, J.: Faster r-cnn: towards real-time object detection with region proposal networks. In: Advances in Neural Information Processing Systems, vol. 28 (2015)
16. Touvron, H., Cord, M., Douze, M., Massa, F., Sablayrolles, A., Jégou, H.: Training data-efficient image transformers and distillation through attention. In: International Conference on Machine Learning, pp. 10347–10357. PMLR (2021)
17. Yap, M.H. et al.: Analysis towards classification of infection and ischaemia of diabetic foot ulcers. In: 2021 IEEE EMBS International Conference on Biomedical and Health Informatics (BHI), pp. 1–4. IEEE (2021)

18. Wang, G., Li, W., Ourselin, S., Vercauteren, T.: Automatic brain tumor segmentation using convolutional neural networks with test-time augmentation. In: Crimi, A., Bakas, S., Kuijf, H., Keyvan, F., Reyes, M., van Walsum, T. (eds.) BrainLes 2018. LNCS, vol. 11384, pp. 61–72. Springer, Cham (2019). https://doi.org/10.1007/978-3-030-11726-9_6
19. Woo, S., Park, J., Lee, J.Y., Kweon, I.S.: Cbam: convolutional block attention module. In: Proceedings of the European Conference on Computer Vision, (ECCV), pp. 3–19 (2018)
20. Zhang, Y., Liu, H., Hu, Q.: TransFuse: fusing transformers and CNNs for medical image segmentation. In: de Bruijne, M., et al. (eds.) MICCAI 2021. LNCS, vol. 12901, pp. 14–24. Springer, Cham (2021). https://doi.org/10.1007/978-3-030-87193-2_2

Unconditionally Generated and Pseudo-Labeled Synthetic Images for Diabetic Foot Ulcer Segmentation Dataset Extension

Raphael Brüngel[1,2,3]($\boxtimes$) (iD), Sven Koitka[3,4] (iD), and Christoph M. Friedrich[1,2] (iD)

[1] Department of Computer Science, University of Applied Sciences and Arts
Dortmund (FH Dortmund), Dortmund, Germany
raphael.bruengel@fh-dortmund.de
[2] Institute for Medical Informatics, Biometry and Epidemiology (IMIBE), University
Hospital Essen, Essen, Germany
[3] Institute for Artificial Intelligence in Medicine (IKIM), University Hospital Essen,
Essen, Germany
[4] Institute of Diagnostic and Interventional Radiology and Neuroradiology,
University Hospital Essen, Essen, Germany

Abstract. The diabetic foot syndrome is a long-term complication of diabetes mellitus. Affected persons are prone to acquisition of deep tissue injuries due to neuropathy-related sensory impairment. When not detected early, Diabetic Foot Ulcers (DFUs) can manifest. Impaired healing capabilities, given vascular damage, and wound bed colonization promote chronification. Subtle changes in the wound bed indicate complications demanding intervention, otherwise prolonging need for treatment. Hence, to enable proper healing frequent and detailed monitoring is necessary. Deep learning-based segmentation is a key technology for automated DFU analysis at the point-of-care, enabling fast measurement and comprehension of subtle changes over time. The research at hand investigates an approach on semantic DFU segmentation, developed during participation in the DFU Challenge (DFUC) 2022. It involves a large ensemble of 25 models of a Feature Pyramid Network with an SE-ResNeXt101-32x4d backbone. Models were trained on an extended training set, enriched to three times its original size via pseudo-labeled synthetic images, generated via the data-efficient unconditional Style-GAN2+ADA. Segmentation performance achieved by challenge submissions is reported. Further, results of synthetic image generation are presented, achieving notably good quality. Results show that the approach achieved competing results, yet overfitting to the synthetic extension was observed. A critical discussion addresses method potentials and risks, points out limitations, and suggests improvements. The work concludes that training set extension with unconditionally generated and pseudo-labeled synthetic images can be achieved rather effortlessly, but increases computational costs and experiment complexity.

Keywords: Diabetic foot ulcers · Wound segmentation · Generative adversarial networks · Synthetic images · Pseudo-labeling

© The Author(s), under exclusive license to Springer Nature Switzerland AG 2023
M. H. Yap et al. (Eds.): DFUC 2022, LNCS 13797, pp. 65–79, 2023.
https://doi.org/10.1007/978-3-031-26354-5_6

1   Introduction

Prevalence estimations for diabetes mellitus from 2021 [29] predicted an amount of 536.6 million cases (10.5%) in the group of 20–79 year olds, which is expected to rise to 783.2 million cases (12.2%) until 2045. A long-term complication of the disease is the diabetic foot syndrome, involving neuropathy and/or vascular damage. Resulting sensory impairment makes affected persons prone to acquisition of ulcers, as deep tissue injuries due to pressure, sheering, or other sources of stress are not recognized. Chronification of injuries is likely due to impaired wound healing capabilities [6] and common colonization-related infections [28]. In 2016, the prevalence of such Diabetic Foot Ulcers (DFUs) was estimated to be approx. 6.3% in the population of diabetics [39].

To enable proper healing and to avoid prolonged treatment or severe complications that demand partial amputation, complex wound management for DFUs relies on frequent and careful monitoring. Subtle changes in the wound dynamic can indicate imminent deterioration of the wound state. However, adequate wound documentation is hard to achieve in the stressful everyday work of overburdened caregivers when implemented manually. Thus, the need for reliable tools enabling partial automation of the wound documentation process is high. Deep learning-based technologies are a key in this matter. Applications for the point-of-care can accelerate time-consuming tasks and support decision making by providing automated analyses for image-based wound documentation, e.g.: Wound area detection [35], wound state assessment [8], wound healing progression measurement [34], or even holistic wound tissue composition breakdown [37]. Exploring performance, optimization potential, and limitations of respective technologies is a necessity to improve and create novel applications.

The work at hand elaborates a contribution to the DFUC2022 [36] on semantic segmentation of DFUs in 2D photographs for medical documentation. The approach involves a segmentation model ensemble that achieved 6th place, using a Feature Pyramid Network (FPN) architecture [18] with an SE-ResNeXt101-32x4d [11,33] backbone, implemented via Segmentation Models PyTorch[1] (SMP). Individual models were trained on an extended training set, extended with synthetic images to three times its original size. Synthetic images were generated by a generation model based on StyleGAN2+ADA [13], pseudo-labeled by another baseline model ensemble trained on the baseline training set.

A baseline study on the DFUC2022 dataset was conducted by [17]. Related work on other public and private DFU segmentation datasets was lately conducted by numerous studies, taking [9,24,27,31] into account as notable and recent examples of stand-alone and combined approaches. Another recently conducted challenge which focused on DFU segmentation was the Foot Ulcer Segmentation Challenge (FUSC) 2021 [32], with [19] as winning approach. In regards to synthetic image generation, [26,38] investigated on general wound, but not

---

[1] Segmentation Models PyTorch: https://github.com/qubvel/segmentation_models.pytorch, accessed 2022-09-10.

specifically on DFU representations. Effects of DFU-specific representations on infection and ischaemia classification performance were investigated by [1].

The following sections cover elaborations on the used dataset and methods, the approach on DFU segmentation as well as dataset extension via synthetic image generation and pseudo-labeling, achieved segmentation and generation results, a discussion on findings and limitations, and conclusions.

2   Dataset and Methods

In the following sections, the DFUC2022 dataset, the used FPN- and SE-ResNeXt101-32x4d-based segmentation network implemented via SMP, and StyleGAN2+ADA for synthetic image generation are described. Additionally, details on the experimental environment are given.

2.1   Diabetic Foot Ulcer Challenge 2022 Dataset

The DFUC2022 dataset [17] addresses single-class semantic segmentation of DFUs in 2D photographs for wound documentation. Images were gathered at the Lancashire Teaching Hospitals[2] at the point-of-care in a non-laboratory environment [4]. Hence, images show typical flaws, e.g., poor lighting and focus, motion blurring, and reflections. DFU delineation for images was created by experts, involving podiatrists and medical consultants. Dataset ground truth was then compiled applying the Snakes [15] active contours algorithm.

The DFUC2020 dataset comprises 4000 images, equally divided into training and testing sets. Ground truth is available solely for the training set. During the challenge a subset of the testing set was used as a validation set, comprising 200 images. Images are provided as Joint Photographic Experts Group (JPEG) files in Red, Green, Blue (RGB) color format with a resolution of 640×480 px. The compression rate of sets differs, the training set shows 70%, the testing set 95%. Ground truth of the training set is provided as masks in the Portable Network Graphics (PNG) format with the same resolution as corresponding images. It comprises 2304 DFU instances with a pixel area range from 0.4‰ up to 35.04% of the whole image pixel area. [17]. DFU instances are generally depicted rather small, as 89.15% of these take <5% of the image pixel size [17]. It should be noted that masks do not display binary but grayscale values. Border regions of instances show a variety of intensity values, presumably due to the applied algorithmic processing. Hence, the exact instance borders are not obvious and the undisclosed evaluation method implementations may be disputable.

Validation and testing set predictions were evaluated via submissions to the DFUC2022 Grand Challenge portal[3]. The main evaluation metrics were the Dice Coefficient [5], the Jaccard Index [12] also known as Intersection over Union (IoU), the False Negative Error (FNE), and the False Positive Error (FPE).

---

[2] Lancashire Teaching Hospitals: https://www.lancsteachinghospitals.nhs.uk/, access 2022-09-10.

[3] DFUC 2022 portal: https://dfuc2022.grand-challenge.org/, access 2022-09-10.

2.2   Semantic Segmentation via Segmentation Models PyTorch

The segmentation approach was implemented via SMP in version 0.2.0., a PyTorch [21]-based framework featuring a variety of architectures, backbones with pre-trained weights, loss function implementations, and metric calculations to enable implementation of unified segmentation pipelines. The FPN [18] architecture was used for semantic segmentation, originally developed for object detection. It predicts at different scales, exploiting the multi-scale pyramidal build-up of deep networks for feature pyramid construction. The top-down structure has lateral connections, with high-level feature maps built at all scales. The used backbone was an SE-ResNeXt101-32x4d, a ResNeXt [33] variant extended with a Squeeze-and-Excitation (SE) [11] module. ResNeXt has a homogeneous design, mainly consisting of repeatedly chained building blocks, aggregating transformations with identical topology. The transformations set size is understood as "cardinality", seen as crucial additional dimension beside depth and width. The SE module additionally focuses on channel relationship to further performance improvement. It recalibrates channel-wise feature responses adaptively, modeling channel interdependencies.

2.3   Image Synthesis via StyleGAN2+ADA

For image synthesis, a generation model was created using StyleGAN2+ADA [13], an unconditional Generative Adversarial Network (GAN) [7] of the Style-GAN [14] family. It implements an Adaptive Discriminator Augmentation (ADA) mechanism to achieve data efficiency, allowing training on small datasets without the problem of early overfitting. It further prevents leakage of augmented image distribution properties to the generated images. The ADA pipeline can be customized, enabling specific tiers of augmentations to be applied during training, comprising [13]: Pixel blitting (x-flips, 90° rotations, translations), general geometric transformations (isotropic scaling, arbitrary rotation, anisotropic scaling, fractional translation), color transformations (brightness, contrast, luma flip, hue rotation, saturation), image-space filtering (four equally sized frequency bands in $[0, \pi]$), and image-space corruptions (additive RGB noise, cutout). To evaluate the quality of generated synthetic images, StyleGAN2+ADA implements the Frechét Inception Distance (FID) as suggested by [10], comparing $50,000$ generated images against the full set of training data.

2.4   Experimental Environment

Experiments were conducted on NVIDIA® V100[4] Graphical Processing Units (GPUs) with 16 GB memory. These were part of an NVIDIA® DGX-1[5], a supercomputer specialized in accelerating deep learning. The operating system was Ubuntu Linux[6] in version 20.04.2 LTS (Focal Fossa), using kernel

---

[4] V100: https://www.nvidia.com/en-us/data-center/v100/, access 2022-09-10.

[5] DGX-1: https://www.nvidia.com/en-us/data-center/dgx-1/, access 2022-09-10.

[6] Ubuntu Linux: https://ubuntu.com/, access 2022-09-10.

version `5.4.0-74-generic`, driver version `450.119.04`, and Compute Unified Device Architecture (CUDA) version `11.0`. The execution environment was an NVIDIA®-optimized `Docker` [20] container engine. For experiments relying on SMP, a Deepo[7] image for a quick setup was used, operating on a single GPU. Experiments with StyleGAN2+ADA were performed in the Docker image provided by the maintainers, operating on four GPUs. Used software packages, referenced in the following description on the approach, are listed in Table 1.

Table 1. Used software packages with version, website, and reference.

Software package	Version	Website, access 2022-09-10	Ref.
Python	3.6.9	https://www.python.org/	–
PyTorch	1.8.1	https://pytorch.org/	[21]
Segmentation Models PyTorch	0.2.0	https://github.com/qubvel/segmentation_models.pytorch	–
Albumentations	1.1.0	https://github.com/albumentations-team/albumentations	[3]
OpenCV	4.2.0.34	https://opencv.org/	[2]
scikit-learn	0.24.2	https://scikit-learn.org/	[22]
ResNeXt	–[a]	https://github.com/facebookresearch/ResNeXt	[33]
Squeeze-and-Excitation Networks	–[a]	https://github.com/hujie-frank/SENet	[11]
StyleGAN2+ADA	–	https://github.com/NVlabs/stylegan2-ada-pytorch	[13]
Docker	19.03.15	https://www.docker.com/	[20]
NVIDIA Container Toolkit	2.5.0	https://github.com/NVIDIA/nvidia-docker	–
Deepo Image	–	https://github.com/ufoym/deepo	–
StyleGAN2+ADA Image	–	https://github.com/NVlabs/stylegan2-ada-pytorch	–
fdupes	2.1.2	https://github.com/adrianlopezroche/fdupes	–
imagemagick	7.1.0-35	https://imagemagick.org/	–

[a] The `se_resnext101_32x4d` version of Segmentation Models Pytorch was used.

3   Approach

The approach involved three stages: (i) training and ensembling of baseline segmentation models, (ii) training of a generation model and training dataset extension, and (iii) training and ensembling of extended segmentation models. Details on applied pre-processing, model training with used hyperparameters and augmentations, aspects of synthetic image generation and pseudo-labeling, and inference with applied post-processing are elaborated in the following sections.

3.1   Pre-processing

An explorative analysis of the training set ground truth revealed that masks were not binary, but held intensity values between $[0, 255]$ at the borders of Regions of Interest (ROIs). Masks were hence binarized via `imagemagick`, using a threshold of ≥ 128 to neither involve too many, nor too few border pixels. A duplicate image analysis was performed via `fdupes`, yielding a set of 39 images in the training dataset that were identical at the binary level. However, the ground

---

[7] Deepo: https://github.com/ufoym/deepo, access 2022-09-10.

truth of each pair showed slight to moderate differences. For each pair a single image was kept, merging the respective ground truth of both via a logical OR operation, implemented via `OpenCV` [2]. The resulting thresholded and cleansed training dataset, in the following referred to as "clean", contained 1961 samples with 2262 DFU instances. Finally, the clean training dataset was prepared to be used within SMP. Images were converted to the PNG format via `imagemagick`, intensity value of masks were mapped from 255 to 1 via `OpenCV`.

3.2   Augmentations

A consistent augmentation pipeline was used throughout all experiments, implemented via `Albumentations` [3], details are elaborated in Table 2. Images were first randomly cropped to a patch of 352×352 px. Following augmentations then were applied with a probability of $p = 0.5$, involving: (i) flips, (ii) a combination of shifts, scaling, and rotation, (iii) either grid distortion or elastic transformation, (iv) either Contrast Limited Adaptive Histogram Equalization (CLAHE), random gamma, or random brightness and contrast, (v) either sharpening, blurring, or motion blurring, and (vi) gaussian noise.

Table 2. Augmentation pipeline, executed in order from top to bottom.

Type	Augmentation	Configuration	Probability
Geometric	RandomCrop	height/width=352	$p = 1.0$
	Flip	Default settings	$p = 0.5$
	ShiftScaleRotate	Default settings	$p = 0.5$
Distortion	GridDistortion	Default settings	$p = 0.5$ for one
	ElasticTransform	Default settings	
Contrast, brightness	CLAHE	Default settings	$p = 0.5$ for one
	RandomGamma	Default settings	
	RandomBrightnessContrast	Default settings	
Blurring, sharpening	Sharpen	Default settings	$p = 0.5$ for one
	Blur	blur_limit=8	
	MotionBlur	blur_limit=8	
Noise	GaussianNoise	Default settings	$p = 0.5$

3.3   Baseline Model Training and Ensemble

To validate model configurations a stratified 5-fold cross-validation was used, implemented via `scikit-learn`, taking into account the DFU instance areas per image. As multiple images in the training set could belong to single subjects and no subject metadata is provided with the DFUC 2022 dataset, fold splits on subject level were not possible. Hence, independence of created validation sets of fold splits is at question, potentially leading to overestimation of model performances during the cross-validation.

Five models were trained on the clean baseline dataset, using identical hyper-parameters and settings, shown in Table 3. For each of these, weights of the last five epochs were saved, resulting in a total of 25 snapshots. These were then combined to an ensemble in the style of Polyak–Ruppert averaging [23, 25].

3.4   Synthetic Image Generation and Dataset Extension

Synthetic representations of DFUs were established training a StyleGAN2+ADA [13] generation model on the pre-processed training set. A dataset preparation tool provided alongside was used to create a fitting dataset, using 512×512 px center crops of training set images. Training was performed for 1000 steps with a batch size of 32, using the default 512 px configuration with pre-trained weights from the Flickr Faces HQ [16] dataset. The default ADA settings were used, additionally activating mirroring for amplification. An FID minimum of 19.09 was identified at 880 steps, hence the respective model checkpoint weights were used for generation of $4,000$ synthetic images, using the seeds $0 - 3999$.

The priorly established baseline segmentation model ensemble was then used to infer predictions on generated synthetic images, using a confidence threshold of 50%. Simple post-processing was applied to resulting pseudo-labels, further described in Sect. 3.6. As generated images did not show compression artifacts like original DFUC2022 dataset images, these were temporarily converted to JPEG images via `imagemagick` with a quality of 95 % and a sampling factor of $2 : 2 : 0$. These setting were oriented towards testing set properties.

For dataset extension, respective generated images were added to the baseline training set. Fold training sets of the established stratified cross-validation setup were extended as well. This was done to observe influence of synthetic images on the performance of extended segmentation models, validated during training on fold validation sets consisting solely of original baseline samples.

3.5   Extended Model Training and Ensemble

Training of extended segmentation models took place under the same conditions as for baseline models. The extended stratified cross-validation was performed to validate training configurations. Five models were trained using the same settings, hyperparameters, and configurations as displayed in Table 3. Again, weights of the last five epochs during each model training process were persisted, forming an averaged ensemble from the 25 resulting weights.

3.6   Inference and Post-processing

For predictions of the baseline segmentation model ensemble for pseudo-labels, and predictions from the extended segmentation model ensemble used for final submissions to the DFUC2022, a confidence threshold of 50 % was set. Post-processing of predictions involved methods of classic image processing, using `OpenCV`. A contour finding algorithm [30] was used to detect single segmentation instances. Instances were measured, removing those with a percentage area

Table 3. Hyperparameters, augmentations, and training set variant.

Parameter	Baseline settings	Extended settings
Architecture	FPN	FPN
Backbone	SE-ResNeXt101-32x4d	SE-ResNeXt101-32x4d
Pre-training weights	ImageNet	ImageNet
Activation function	sigmoid	sigmoid
Loss function	Dice	Dice
Optimizer	Adam	Adam
Initial learning rate	0.0001	0.0001
Dropped learning rate	0.00005, 0.00001	0.00001
Epochs (dropped)	150 (100, 135)	150 (120)
Batch size	24	24
Number of GPUs	1	1
Augmentations	see Table 2	see Table 2
Dataset	Clean baseline (1961 images)	Extended (1961+4000 images)

of $< 1\%_0$ in regards to the image size. Predictions with just one instance were excluded from this filtering procedure. Following, an opening was applied to remove potential small artifacts from the contour removal step as well as fringes, using a 2×2 square kernel. The described procedure was applied for all predictions of the baseline segmentation model ensemble comprising validation set predictions and pseudo-labeling of synthetic images. Whether it was applied for predictions of the extended segmentation models ensemble for testing set predictions is indicated in the reporting of results.

4   Results

Results of the baseline and extended segmentation model ensemble are reported in the following. Illustrations on the quality of generated synthetic images and inferred pseudo-labels are given as well.

4.1   Baseline and Extended Segmentation Ensemble Performance

Results of experiments are presented in Table 4, showing performance of the baseline and extended segmentation ensemble on the validation and testing set. Two final submissions for the testing set were prepared based on the described approach, one without and one with post-processing involved. A subsequent evaluation of baseline ensemble performance is reported as well.

During the validation phase, all submissions for presented approach involved post-processing. It was applied consistently due to the early observation of strong beneficial effects on scores. While the extended ensemble performed notably better on the validation set than the baseline ensemble, this was not the case for the

Table 4. Results for baseline and extended segmentation model ensembles, best results per challenge phase are highlighted.

Configuration	Post-processing	Dice ↑	Jaccard ↑	FNE ↓	FPE ↓
Mid-challenge: Submissions for validation set					
Baseline ensemble	Yes[a]	0.6895	0.5880	0.2693	0.2493
Extended ensemble	Yes[a]	**0.6971**	**0.5974**	**0.2578**	**0.2466**
End-challenge: Final submissions for testing set					
Extended ensemble	No	**0.7169**	**0.6130**	**0.2453**	**0.2145**
Extended ensemble	Yes	0.7136	0.6086	0.2470	0.2195
Post-challenge: Submissions for testing set (subsequent evaluation)					
Baseline ensemble	No	**0.7211**	**0.6167**	**0.2462**	**0.2064**
Baseline ensemble	Yes	0.7178	0.6125	0.2490	0.2116

[a] During the validation phase only submissions with post-processing were prepared.

testing set, observed in subsequent evaluations. The same applied for the usefulness of the post-processing for which harmful effects were revealed. Beside reported downsides, predicted segmentations of both, baseline and extended ensemble, were visually coherent. Malformed toe nails were occasionally the cause of false-positive predictions. Given macerations, usually a thin band of macerated skin at the wound bed border was accounted to the predicted region of interest.

4.2 Synthetic Images and Pseudo-Labels for Dataset Extension

The generation model allowed to generate a broad variety of qualitatively good synthetic images, a palette of randomly generated examples is shown in Fig. 1. Backgrounds feature well-represented blankets used in the clinical environment and usually do not mix with body parts, yet occasionally with other background items. Medical gloves are present in some scenes. Body parts with none, single, and multiple DFU representations were generated, showing typical accompanying symptoms with a high level of detail. E.g., hyperkeratosis, macerations, ischemic areas, and reddish surroundings indicating infections are present in different grades of manifestations. Due to numerous malformed or partially amputated body parts in original images, generated body part representations tend to not show proportionally correct anatomy of the foot. Surprisingly, representations with six or more toe-like structures are common. Some representations display elongated body parts without complex anatomy. DFU representations show a variety of sizes, shapes, and tissues mix representations. Sizes and shapes vary from small, elliptical, and shallow representations, reminding of early-stage pressure injuries, to large, complex, and deep representations, reminding of complication-ridden chronic wounds. Borders vary from well demarcated edges to fringy areas with weak partial epithelialization. Wound beds can feature isle-like partial epithelialization as well.

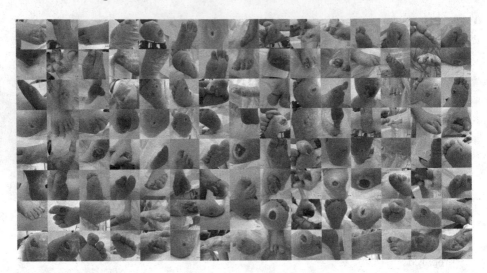

Fig. 1. Overview of synthetic image output from the trained generation model.

Examples of created pseudo-labels for generated synthetic images are presented in Fig. 2, showing a visually coherent delineation of DFU areas. Shown examples are representative for most created pseudo-labels. However, generated representations of malformed toe nails were occasionally segmented as well, which is also a common issue when inferring predictions on real images.

5 Discussion

The following discussion addresses experiments and achieved results. Limitations are pointed out as well, suggesting optimizations and alternative approaches.

5.1 Segmentation Performance and Post-processing

The extended ensemble without post-processing achieved place 6 of 17 with a Dice score of 0.7169, exceeding the threshold of ≥ 0.7 set by the maintainers to be accounted as top approach. This was accomplished by place 1 with a Dice score of 0.7287 to place 6, the presented approach.

However, post-challenge evaluations show that the highest performance within the approach was achieved by the baseline ensemble without post-processing. While both, synthetic training set extension and post-processing, increased ensemble performance on the validation set, their arrangement appeared to have a harmful effects on the testing set predictions. The post-processing procedure was initially drafted to eliminate tiny instances with an area of $<1\%_0$ image size, perceived as artifacts in the majority of cases. Yet, tiny instances seem to be more common in the testing set ground truth. Further elaborations on the training set extension is given in the following section.

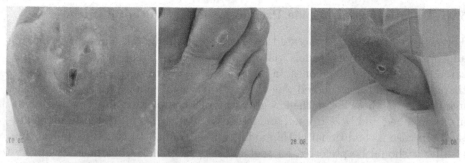

(a) Small-sized instances of DFU representations.

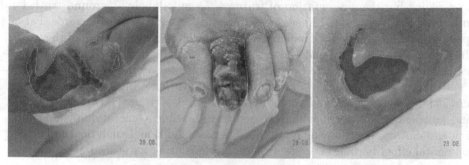

(b) Medium-sized instances of DFU representations.

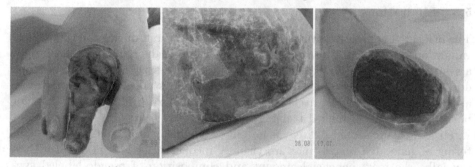

(c) Large-sized instances of DFU representations.

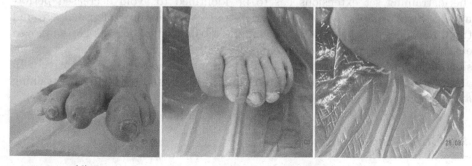

(d) Instanceless representations of presumably uninjured feet.

Fig. 2. Pseudo-label examples (cyan demarcation) by the baseline ensemble.

5.2   Dataset Extension via Pseudo-Labeled Synthetic Images

The achieved quality of generated images is generally good. However, achieving appropriate generation models via unconditional GANs is usually less complicated than via conditional ones. In fact, attempts to utilize conditional GANs for the approach did not succeed due to very low visual quality and artifacts. The main cause for these lies in the low ground truth complexity, simply discriminating wound bed and background, and lesser data-efficiency of tested implementations. In case a conditional GAN model can be trained to generate appropriate results, it should be preferred for segmentation problems. The benefits of these are direct output control via masks that also serve as appropriate ground truth, and region-based representation of features. Pseudo-labels as ground truth for unconditionally generated images are always disputable in these regards. False-positive and false-negative predictions can occur in the widest sense of the word. Inclusion of these into pseudo-ground truth of an extended dataset can amplify certainty for such, calculated by predicting segmentation models trained on it.

Training set extension via large amounts of synthetic images can be seen as critical due to trade-offs and potential pitfalls. Models trained on extended training sets demand proportionally increased training time. Generation model training and image generation add up to this. Further, additional analytical and developmental effort becomes necessary when implementing fine-tuned approaches.

5.3   Limitations

The presented approach features several limitations. First, during the DFUC2022 validation stage only explorative experiments were conducted. For better insight a systematic comparison of all aspects would be necessary, yet was out of scope due to limited submissions. Grid-search or ablation studies may identify more appropriate architectures, backbones, hyperparameters, and augmentations for both, baseline and extended training sets. Especially in regards to training set extension, the presented strategy is unrefined. Refinement should investigate filtering of generated images that are visually incoherent or lack in quality. Most crucially, the most beneficial amount and composition of pseudo-labeled synthetic training set extensions has to be identified to avoid decreased performance. Sizes, shapes, color histograms, and other features of pseudo-ground truth instances may be analyzed to compile a balanced extension. The employed ensembling strategy is rather unrefined as well. Influence of training set extension on individual models is hard to assess if not evaluated comprehensively. In addition, utilization of large model ensembles hinders applicability, as training and inference time prove costly.

6   Conclusion

The employed large ensemble of FPN models with SE-ResNeXt101-32x4d backbone in its baseline configuration achieves competing results. Beneficial effects of applied post-processing for removal of tiny artifact-like instances during the validation phase proved to be deceptive. The same applies to the vast amount of synthetic training set extension. Yet, data-efficient unconditional GANs such as StyleGAN2+ADA enable generation of realistic DFU representations on the basis of small datasets. Pseudo-labeling these displays an effortless way for arbitrary extension of segmentation datasets. The approach illustrates an alternative to the use of conditional GANs that are more difficult to train on small datasets with low complexity ground truth. However, careful investigations on appropriate amounts, quality of generated images, and created pseudo-ground truth are advised. Future work will address these aspects to unveil viable practices.

Acknowledgments. Raphael Brüngel was partially funded by a PhD grant from the University of Applied Sciences and Arts Dortmund, Dortmund, Germany.

References

1. Bloch, L., Brüngel, R., Friedrich, C.M.: Boosting EfficientNets ensemble performance via pseudo-labels and synthetic images by pix2pixHD for infection and Ischaemia classification in diabetic foot ulcers. In: Yap, M.H., Cassidy, B., Kendrick, C. (eds.) DFUC 2021. LNCS, vol. 13183, pp. 30–49. Springer, Cham (2022). https://doi.org/10.1007/978-3-030-94907-5_3
2. Bradski, G.: The OpenCV library. Dr. Dobb's J. Softw. Tools **25**, 122–125 (2000)
3. Buslaev, A., Iglovikov, V.I., Khvedchenya, E., Parinov, A., Druzhinin, M., Kalinin, A.A.: Albumentations: fast and flexible image augmentations. Information **11**(2), 125 (2020). https://doi.org/10.3390/info11020125
4. Cassidy, B., et al.: The DFUC 2020 dataset: analysis towards diabetic foot ulcer detection. touchREVIEWS Endocrinol. **17**, 5–11 (2021). https://doi.org/10.17925/EE.2021.17.1.5
5. Dice, L.R.: Measures of the amount of ecologic association between species. Ecology **26**(3), 297–302 (1945). https://doi.org/10.2307/1932409
6. Falanga, V.: Wound healing and its impairment in the diabetic foot. Lancet **366**(9498), 1736–1743 (2005). https://doi.org/10.1016/s0140-6736(05)67700-8
7. Goodfellow, I., et al.: Generative adversarial networks. In: Ghahramani, Z., Welling, M., Cortes, C., Lawrence, N., Weinberger, K. (eds.) Advances in Neural Information Processing Systems. vol. 27. Curran Associates, Inc. (2014)
8. Goyal, M., Reeves, N.D., Rajbhandari, S., Ahmad, N., Wang, C., Yap, M.H.: Recognition of Ischaemia and infection in diabetic foot ulcers: dataset and techniques. Comput. Biol. Med. **117**, 103616 (2020). https://doi.org/10.1016/j.compbiomed.2020.103616
9. Goyal, M., Yap, M.H., Reeves, N.D., Rajbhandari, S., Spragg, J.: Fully convolutional networks for diabetic foot ulcer segmentation. In: 2017 IEEE International Conference on Systems, Man, and Cybernetics (SMC), pp. 618–623 (2017). https://doi.org/10.1109/SMC.2017.8122675
10. Heusel, M., Ramsauer, H., Unterthiner, T., Nessler, B., Hochreiter, S.: GANs trained by a two time-scale update rule converge to a local nash equilibrium. In: Guyon, I., et al. (eds.) Advances in Neural Information Processing Systems, vol. 30. Curran Associates, Inc. (2017)

11. Hu, J., Shen, L., Albanie, S., Sun, G., Wu, E.: Squeeze-and-excitation networks. IEEE Trans. Pattern Anal. Mach. Intell. **42**(8), 2011–2023 (2020). https://doi.org/10.1109/TPAMI.2019.2913372

12. Jaccard, P.: The distribution of the flora in the alpine zone. New Phytol. **11**(2), 37–50 (1912). https://doi.org/10.1111/j.1469-8137.1912.tb05611.x

13. Karras, T., Aittala, M., Hellsten, J., Laine, S., Lehtinen, J., Aila, T.: Training generative adversarial networks with limited data. In: Larochelle, H., Ranzato, M., Hadsell, R., Balcan, M.F., Lin, H. (eds.) Advances in Neural Information Processing Systems (NeurIPS 2020), vol. 33, pp. 12104–12114. Curran Associates, Inc. (2020)

14. Karras, T., Laine, S., Aila, T.: A style-based generator architecture for generative adversarial networks. In: Proceedings of the IEEE/CVF Conference on Computer Vision and Pattern Recognition (CVPR 2019), pp. 4396–4405 (2019). https://doi.org/10.1109/CVPR.2019.00453

15. Kass, M., Witkin, A., Terzopoulos, D.: Snakes: active contour models. Int. J. Comput. Vis. **1**(4), 321–331 (1988). https://doi.org/10.1007/bf00133570

16. Kazemi, V., Sullivan, J.: One millisecond face alignment with an ensemble of regression trees. In: 2014 IEEE Conference on Computer Vision and Pattern Recognition, pp. 1867–1874 (2014). https://doi.org/10.1109/CVPR.2014.241

17. Kendrick, C., et al.: Translating clinical delineation of diabetic foot ulcers into machine interpretable segmentation (2022). https://doi.org/10.48550/ARXIV.2204.11618

18. Lin, T.Y., Dollár, P., Girshick, R., He, K., Hariharan, B., Belongie, S.: Feature pyramid networks for object detection. In: 2017 IEEE Conference on Computer Vision and Pattern Recognition (CVPR), pp. 936–944 (2017). https://doi.org/10.1109/CVPR.2017.106

19. Mahbod, A., Schaefer, G., Ecker, R., Ellinger, I.: Automatic foot ulcer segmentation using an ensemble of convolutional neural networks. In: 2022 26th International Conference on Pattern Recognition (ICPR), pp. 4358–4364. IEEE (2021). https://doi.org/10.48550/ARXIV.2109.01408

20. Merkel, D.: Docker: lightweight Linux containers for consistent development and deployment. Linux J. **239**(2), 2 (2014)

21. Paszke, A., et al.: PyTorch: an imperative style, high-performance deep learning library. In: Wallach, H., Larochelle, H., Beygelzimer, A., d'Alché Buc, F., Fox, E., Garnett, R. (eds.) Advances in Neural Information Processing Systems (NeuriPS 2019), vol. 32, pp. 8024–8035. Curran Associates, Inc. (2019)

22. Pedregosa, F., et al.: Scikit-learn: machine learning in Python. J. Mach. Learn. Res. **12**, 2825–2830 (2011)

23. Polyak, B.: New stochastic approximation type procedures. Avtomatica i Telemekhanika **7**, 98–107 (1990)

24. Rania, N., Douzi, H., Yves, L., Sylvie, T.: Semantic segmentation of diabetic foot ulcer images: dealing with small dataset in DL approaches. In: El Moataz, A., Mammass, D., Mansouri, A., Nouboud, F. (eds.) ICISP 2020. LNCS, vol. 12119, pp. 162–169. Springer, Cham (2020). https://doi.org/10.1007/978-3-030-51935-3_17

25. Ruppert, D.: Efficient estimations from a slowly convergent Robbins-Monro process. Tech. rep. Cornell University Operations Research and Industrial Engineering (1988)

26. Sarp, S., Kuzlu, M., Wilson, E., Guler, O.: WG2AN: synthetic wound image generation using generative adversarial network. J. Eng. **2021**(5), 286–294 (2021). https://doi.org/10.1049/tje2.12033

27. Scebba, G., et al.: Detect-and-segment: a deep learning approach to automate wound image segmentation. Inform. Med. Unlocked **29**, 100884 (2022). https://doi.org/10.1016/j.imu.2022.100884
28. Siddiqui, A.R., Bernstein, J.M.: Chronic wound infection: facts and controversies. Clin. Dermatol. **28**(5), 519–526 (2010). https://doi.org/10.1016/j.clindermatol.2010.03.009
29. Sun, H., et al.: IDF diabetes atlas: global, regional and country-level diabetes prevalence estimates for 2021 and projections for 2045. Diabetes Res. Clin. Pract. **183**, 109119 (2022). https://doi.org/10.1016/j.diabres.2021.109119
30. Suzuki, S., Abe, K.: Topological structural analysis of digitized binary images by border following. Comput. Vis. Graph. Image Process. **30**(1), 32–46 (1985). https://doi.org/10.1016/0734-189X(85)90016-7
31. Wang, C., et al.: Fully automatic wound segmentation with deep convolutional neural networks. Sci. Rep. **10**(1), 1–9 (2020). https://doi.org/10.1038/s41598-020-78799-w
32. Wang, C., et al.: FUSeg: The Foot Ulcer Segmentation Challenge. arXiv preprint arXiv:2201.00414 (2022)
33. Xie, S., Girshick, R., Dollár, P., Tu, Z., He, K.: Aggregated residual transformations for deep neural networks. In: 2017 IEEE Conference on Computer Vision and Pattern Recognition (CVPR), pp. 5987–5995 (2017). https://doi.org/10.1109/CVPR.2017.634
34. Yang, S., et al.: Sequential change of wound calculated by image analysis using a color patch method during a secondary intention healing. PLoS ONE **11**(9), 1–15 (2016). https://doi.org/10.1371/journal.pone.0163092
35. Yap, M.H., et al.: Deep learning in diabetic foot ulcers detection: a comprehensive evaluation. Comput. Biol. Med. **135**, 104596 (2021). https://doi.org/10.1016/j.compbiomed.2021.104596
36. Yap, M.H., et al.: Diabetic foot ulcers grand challenge (2022). https://doi.org/10.5281/zenodo.4575228
37. Zahia, S., Sierra-Sosa, D., Garcia-Zapirain, B., Elmaghraby, A.: Tissue classification and segmentation of pressure injuries using convolutional neural networks. Comput. Methods Programs Biomed. **159**, 51–58 (2018). https://doi.org/10.1016/j.cmpb.2018.02.018
38. Zhang, J., Zhu, E., Guo, X., Chen, H., Yin, J.: Chronic wounds image generator based on deep convolutional generative adversarial networks. In: Li, L., Lu, P., He, K. (eds.) NCTCS 2018. CCIS, vol. 882, pp. 150–158. Springer, Singapore (2018). https://doi.org/10.1007/978-981-13-2712-4_11
39. Zhang, P., Lu, J., Jing, Y., Tang, S., Zhu, D., Bi, Y.: Global epidemiology of diabetic foot ulceration: a systematic review and meta-analysis. Ann. Med. **49**(2), 106–116 (2016). https://doi.org/10.1080/07853890.2016.1231932

Post Challenge Papers

Diabetic Foot Ulcer Segmentation Using Convolutional and Transformer-Based Models

Mariam Hassib[1]([✉]), Maram Ali[1], Amina Mohamed[1], Marwan Torki[1][iD],
and Mohamed Hussein[2][iD]

[1] Faculty of Engineering, Alexandria University, Alexandria, Egypt
`{es-Mariam.Mahmoud2022,eng-maram.attia1722,es-Amina.Ali2022,`
`mtorki}@alexu.edu.eg`
[2] Information Sciences Institute, University of Southern California, Marina del Rey,
CA 90292, USA
`mehussein@isi.edu`

Abstract. Diabetes is a rising global epidemic, it was estimated that in 2017 there are 451 million (aged 18–99 years) people with diabetes worldwide, and it is expected to increase to 693 million by 2045. Diabetic Foot Ulcers (DFU) is a serious disease affecting diabetic patients and can lead to limb amputation, while more serious cases can even lead to death. In an effort to improve patient care, we are taking part in the Diabetic Foot Ulcer Segmentation Challenge 2022 (DFUC2022) competition to design automated computer methods for ulcers segmentation. This paper summarises our proposed method for the DFUC2022 conducted in conjunction with MICCAI 2022. Our experiments are based on convolutional and transformer-based models. The best performing model of our proposed method was the SegFormer model, which achieved a dice coefficient of 69.89% and a Jaccard coefficient of 59.21%. The code and the link to the pre-trained models are available at: https://github.com/Maramattia/Diabetic-Foot-Ulcer-Challenge.git.

Keywords: Diabetic Foot Ulcers · Diabetes · Semantic segmentation · Deep learning · SegFormer · DeepLabV3+ · DFUC2022

1 Introduction

A diabetic foot ulcer is an open sore or wound that occurs in approximately 15% of diabetic patients. It is commonly located on the bottom of the foot. Of those who develop a foot ulcer, 6% will be hospitalized due to infection or other ulcer-related complications [5].

There are three different tasks that are performed to detect anomalies in medical images: classification, object detection, and segmentation. Classification is to recognize the type of anomaly. Object detection is to point out the region of the anomaly. Segmentation is to partition the anomaly to analyze its conditions. The overview of the development of DFU in classification, object detection and segmentation is summarised by Yap et al. [11].

© The Author(s), under exclusive license to Springer Nature Switzerland AG 2023
M. H. Yap et al. (Eds.): DFUC 2022, LNCS 13797, pp. 83–91, 2023.
https://doi.org/10.1007/978-3-031-26354-5_7

This work was done for segmenting foot ulcers in individuals with diabetes based on SegFormer and DeepLabV3+. The purpose of this work is to help in the treatment of diabetic foot ulcer disease. The main task is to classify each pixel in the image with the labels "background" and "ulcer". This work is motivated by the creation of the Diabetic Foot Ulcers Grand Challenge 2022 (DFUC2022) [7,12].

2 Background

DeepLabV3+. DeepLabV3+ employs the encoder-decoder structure where the DeepLabv3 network is used to encode the rich contextual information and a simple yet effective decoder module is adopted to recover the object boundaries. One could also apply the atrous convolution to extract the encoder features at an arbitrary resolution, depending on the available computation resources [3].

SegFormer. SegFormer is a simple and powerful semantic segmentation method that consists of a positional-encoding-free, hierarchical Transformer encoder and a lightweight AllMLP decoder. It avoids common complex designs in previous methods, leading to both high efficiency and performance. SegFormer not only achieves new state-of-the-art results on standard datasets but also shows strong zero-shot robustness [10].

3 Related Work

Fully Convolutional Networks. Previous work on diabetic foot ulcer segmentation by [6] was completed using fully convolutional networks. For the segmentation experiments they used a small dataset with 705 images with an FCN-16s network. They applied 5-fold cross-validation with two-tier transfer learning resulting in a Dice Similarity Coefficient of 0.794 ($\pm$0.104) for segmentation of DFU regions. It shows the potential of the fully convolutional network in the DFU segmentation task.

Ensemble. Recently, at the MICCAI 2021 online only event, Foot Ulcer Segmentation (FUSeg) Challenge, the winning team used an ensemble approach to achieve state-of-the-art performance with dice scores of 92.07% and 88.80%, respectively, and the top-ranked method in the FUSeg challenge leaderboard. They propose an ensemble approach based on two encoder-decoder-based CNN models, namely LinkNet and U-Net, to perform foot ulcer segmentation [8].

4 Dataset

In this work we use the DFUC2022 dataset released in April 2022 [1,12]. The DFUC2022 training set consists of 2000 images with 2304 ulcers, where the smallest ulcer size is 0.04 % of the total image size, and the largest ulcer

Table 1. DFUC2022 dataset distribution.

Dataset split	Total number of images
Train	2000
Validation	200
Test	2000

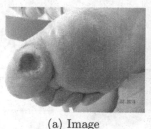

(a) Image   (b) Ground Truth

Fig. 1. An example of (a) an image from the DFUC2022 dataset and (b) the corresponding ground truth mask.

size is 35.04% of the total image size [7]. In the test cases, a small sample of healthy/remission examples are included to test the reliability of trained networks against false positive results [7].

Since there is no annotation provided for the validation dataset, we split the training dataset randomly into training and local validation datasets with a ratio of 95% train and 5% validation for evaluating the accuracy of our model.

5   Proposed Method

5.1   Models

Our solution is an ensemble of two different models: DeepLabV3+ [2] and Seg-Former [10]. First, we explain the details of each model. We implemented the models based on the OpenMMLab Semantic Segmentation Toolbox. The code is available at: https://github.com/open-mmlab/mmsegmentation [4].

DeepLabV3+. DeepLabV3+ as in Fig. 2 is a convolution-based model, One of the challenges in segmenting objects in images using deep convolutional neural networks (DCNNs) is that as the input feature map grows smaller from traversing through the network, information about objects of a smaller scale can be lost [3]. The contribution of this paper [2] DeepLabv3 is the introduction of atrous convolutions or dilated convolutions, to extract more dense features where information is better preserved given objects of varying scale of 2-3.

SegFormer. SegFormer as in Fig. 3 is a simple, efficient yet powerful semantic segmentation framework that unifies Transformers with lightweight multilayer

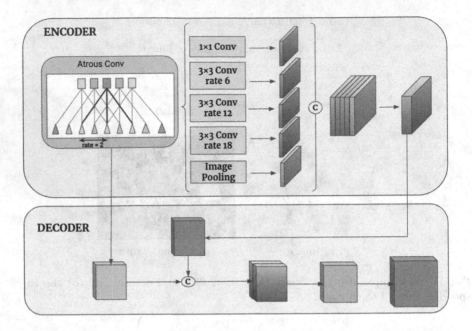

Fig. 2. Illustration of DeepLabV3+ architecture (redrawn from [3]).

perceptron (MLP) decoders. It has two appealing features. First, it comprises a novel hierarchically structured Transformer encoder which outputs multiscale features that does not need positional encoding. Therefore, it avoids the interpolation of positional codes, which leads to decreased performance when the testing resolution differs from training. Second, it avoids complex decoders. SegFormer consists of two main modules: a hierarchical Transformer encoder and a lightweight All-MLP decoder to predict the final mask. Given an image with a size $H \times W \times 3$, first, divides it into patches of size 4×4. Second, these patches are used as input to the hierarchical Transformer encoder to get multilevel features. Finally, these multilevel features are passed to the All-MLP decoder to predict the segmentation mask with a $H/4 \times W/N_{cls}$ resolution [10].

5.2 Training Procedure

The weights of the two models are initialized with the weights trained with ade20k [13]. In the case of the DeepLabV3+ network, we employ a Stochastic Gradient Descent optimizer with a momentum of 0.9, weight decay of 0.0005, and learning rate of 3e−03. Trained DeepLabV3+ for 15k iterations with a batch size 16. In the case of the SegFormer we employ the AdamW optimizer with the betas (0.9, 0.999), weight decay 0.01, and learning rate of 6e−05. We trained the SegFormer for 10k iterations with batch size 16. A single GPU on Google Colab Pro was used for training, which was completed in approximately 24 h.

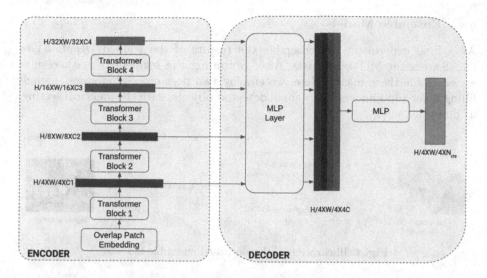

Fig. 3. Illustration of the SegFormer architecture (redrawn from [10]).

Data Augmentation. We apply several data augmentation techniques for achieving robust training with limited data. Specifically, we apply the data augmentation shown in Table 2.

Table 2. Data augmentation used for training in our experiments.

Augmentation type	Probability range
Resize	[0.5, 2.0]
Random crop	0.75
Random flip	0.5
Brightness	32
Contrast	[0.5, 1.5]

5.3   Testing Procedure

At test time, we employ a test-time augmentation scheme using the same strategy as our training augmentation. That is to say, augment the input test with multiple scales, and apply random flips and different photometric distortions. For SegFormer we add an extra augmentation method which is Resize-To-Multiple of 32. It improves SegFormer by 0.5–1.0 mIoU. The original implementation of the SegFormer Model uses different test pipelines and image ratios in ms+flip. In OpenMMLab Semantic Segmentation Toolbox, they replace Aligned-Resize in the original implementation to Resize + Resize-To-Multiple. Finally, we merge the predictions against different images.

5.4   Ensemble Models

As a final experiment, we ensemble the results of the trained DeepLabV3+ and SegFormer MiT-b5 models. After obtaining the segmented mask results, we ensemble these masks. Here, we employ two post-processing steps, namely filling holes and removing very small detected objects, with the identical settings as described in [9].

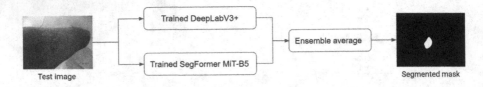

Fig. 4. Illustration of the proposed ensemble models.

Table 3. DFUC2022 leaderboard test results for our proposed method. FNE: False Negative Error; FPE: False Positive Error.

Method	Dice coefficient	Jaccard coefficient	FNE	FPE
DeepLabV3+	0.6972	0.5906	0.2855	0.1998
SegFormer MiT-B5	**0.6986**	**0.5921**	**0.2778**	0.2064
Ensemble	0.6961	0.5887	0.3011	**0.1823**

6   Results

After fine-tuning DeepLabV3+ and SegFormer MiT-B5, we tested the models which showed the best performance on our local validation set. We summarize the test submission results on the leaderboard at Table 3. In our experiments, we achieved 69.86%, 69.72%, and 69.61% of dice coefficient by using SegFomer MiT-B5, DeepLabV3+, and ensemble, respectively. SegFormer MiT-B5 performance is better than DeepLabV3+ on the test dataset. The ensemble of both models did not increase the dice coefficient score, it increased the false negative error.

Post Challenge Analysis. After the release of challenge results, we performed additional analysis on the test-time augmentation (TTA), to see if it has an influence on the final ensemble. From Table 4, it seems that even without using TTA, the ensemble did not improve in terms of the Dice Coefficient. Therefore, test-time augmentation improves the predicted masks and has a better effect on the SegFormer model than the DeepLabV3+ model.

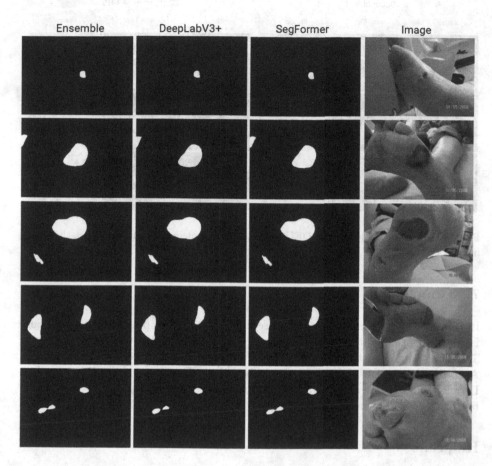

| Ensemble | DeepLabV3+ | SegFormer | Image |

Fig. 5. Predicted masks on the DFUC2022 test dataset.

Table 4. Test-time augmentation analysis. FNE: False Negative Error; FPE: False Positive Error.

Method	TTA	Dice coefficient	Jaccard coefficient	FNE	FPE
DeepLabV3+		0.6904	0.5862	0.2789	0.2188
	✓	0.6972	0.5906	0.2855	0.1998
SegFormer MiT-B5		0.6894	0.5803	0.2723	0.2339
	✓	**0.6986**	**0.5921**	0.2778	0.2064
Ensemble		0.6870	0.5787	0.3128	0.1813
	✓	0.6961	0.5887	0.3011	0.1823

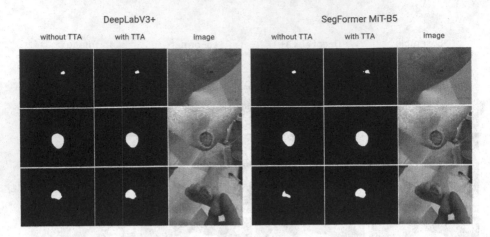

Fig. 6. Predicted masks on the DFUC2022test set showing the effect of test-time augmentation.

7 Conclusion

In this work, we have proposed an ensemble between two powerful models - SegFormer and DeepLabV3+. Our approach did not increase the dice coefficient. Test-time augmentation shows great improvements on the predicted masks. According to the DFUC2022 test leaderboard, SegFormer performed better than DeeplabV3+. However, on the validation leaderboard, DeeplabV3+ performed better than SegFormer.

References

1. Cassidy, B., et al.: The DFUC 2020 dataset: analysis towards diabetic foot ulcer detection. touchREVIEWS in Endocrinology **17**, 5–11 (2021). https://doi.org/10.17925/EE.2021.17.1.5, https://www.touchendocrinology.com/diabetes/journal-articles/the-dfuc-2020-dataset-analysis-towards-diabetic-foot-ulcer-detection/1
2. Chen, L.C., Papandreou, G., Schroff, F., Adam, H.: Rethinking atrous convolution for semantic image segmentation. arXiv preprint. arXiv:1706.05587 (2017)
3. Chen, L.-C., Zhu, Y., Papandreou, G., Schroff, F., Adam, H.: Encoder-decoder with atrous separable convolution for semantic image segmentation. In: Ferrari, V., Hebert, M., Sminchisescu, C., Weiss, Y. (eds.) ECCV 2018. LNCS, vol. 11211, pp. 833–851. Springer, Cham (2018). https://doi.org/10.1007/978-3-030-01234-2_49
4. Contributors, M.: MMSegmentation: openmmlab semantic segmentation toolbox and benchmark. https://github.com/open-mmlab/mmsegmentation (2020)
5. Goyal, M., Reeves, N.D., Rajbhandari, S., Yap, M.H.: Robust methods for real-time diabetic foot ulcer detection and localization on mobile devices. IEEE J. Biomed. Health Inform. **23**(4), 1730–1741 (2018)
6. Goyal, M., Yap, M.H., Reeves, N.D., Rajbhandari, S., Spragg, J.: Fully convolutional networks for diabetic foot ulcer segmentation. In: 2017 IEEE International Conference on Systems, Man, and Cybernetics (SMC), pp. 618–623. IEEE (2017)

7. Kendrick, C., et al.: Translating clinical delineation of diabetic foot ulcers into machine interpretable segmentation. arXiv preprint. arXiv:2204.11618 (2022)
8. Mahbod, A., Ecker, R., Ellinger, I.: Automatic foot ulcer segmentation using an ensemble of convolutional neural networks. arXiv preprint. arXiv:2109.01408 (2021)
9. Wang, C., Anisuzzaman, D., Williamson, V., Dhar, M.K., Rostami, B., Niezgoda, J., Gopalakrishnan, S., Yu, Z.: Fully automatic wound segmentation with deep convolutional neural networks. Sci. Rep. **10**(1), 1–9 (2020)
10. Xie, E., Wang, W., Yu, Z., Anandkumar, A., Alvarez, J.M., Luo, P.: Segformer: simple and efficient design for semantic segmentation with transformers. In: Advances in Neural Information Processing Systems, vol. 34, pp. 12077–12090 (2021)
11. Yap, M.H., Kendrick, C., Reeves, N.D., Goyal, M., Pappachan, J.M., Cassidy, B.: Development of diabetic foot ulcer datasets: an overview. In: Yap, M.H., Cassidy, B., Kendrick, C. (eds.) DFUC 2021. LNCS, vol. 13183, pp. 1–18. Springer, Cham (2022). https://doi.org/10.1007/978-3-030-94907-5_1
12. Yap, M.H., et al.: Diabetic foot ulcers grand challenge 2022 (2021). https://doi.org/10.5281/zenodo.6389665
13. Zhou, B., Zhao, H., Puig, X., Fidler, S., Barriuso, A., Torralba, A.: Scene parsing through ade20k dataset. In: Proceedings of the IEEE Conference on Computer Vision and Pattern Recognition, pp. 633–641 (2017)

Refined Mixup Augmentation
for Diabetic Foot Ulcer Segmentation

David Jozef Hresko[1]([⊠]) [iD], Jakub Vereb[1] [iD], Valentin Krigovsky[1] [iD],
Michala Gayova[2] [iD], and Peter Drotar[1] [iD]

[1] IISlab, Technical University of Kosice, Kosice, Slovakia
david.jozef.hresko@tuke.sk
[2] Department of Burns and Reconstructive Surgery,
PJ Safarik University and Hospital Agel, Kosice-Saca, Slovakia

Abstract. Diabetic foot syndrome is one of the chronic diabetic complications that present itself as foot ulceration, which in many cases leads
to limb amputation. Moreover, it is linked to the high percentage of post-
amputation mortality within the period of five years. Thus it is crucial to
diagnose and plan careful treatment properly in the early stages. Diagnosis is often time-consuming and requires a skilled clinician who can differ-
entiate between similarities between diabetes-related and other types of
ulcers by evaluating their morphology and location. To mitigate diagnostic issues, we propose an improved version of the nnU-Net architecture
with residual short skip connections in the encoder part and additional
mixup augmentation as the preprocessing step. The obtained results on
the DFUC2022 challenge dataset show that our improvements can boost
overall performance for ulcer segmentation tasks, even in scenarios where
targeted structures are heterogeneous and under high imbalance conditions in the evaluated dataset. With our approach we achieved 9th place
with a Dice score of 0.6975.

Keywords: nnU-Net · Mixup augmentation · Skip connections ·
Segmentation · Diabetic ulcer

1 Introduction

Diabetes is a metabolic disease characterized by high blood sugar levels over
a prolonged period of time. Diabetes is already considered a pandemic disease,
and with time, its prevalence is expected only to increase (from 536 million in
2021 to 783 million in 2045) [10]. Even though the diagnosis is straightforward,
it frequently happens late and already in the stage of chronic complications
[8]. Diabetic foot syndrome (DFS) is one of the chronic diabetic complications
that presents as foot ulceration. Despite intensive treatment, the affected limb
is amputated within 6–18 months after the first evaluation in up to a quarter of
patients, while the 5-year post-amputation mortality is as high as 50% [5]. On the
other side, early diagnosis and careful treatment can prevent limb amputation

M. H. Yap et al. (Eds.): DFUC 2022, LNCS 13797, pp. 92–100, 2023.
https://doi.org/10.1007/978-3-031-26354-5_8

and ulcer-related mortality. Therefore, careful treatment must not be delayed due to misdiagnosis.

In recent years, convolutional neural networks (CNN) penetrated almost every aspect of medical imaging. Whether it is classification, segmentation, or registration of medical images, CNN offers performance close to the human specialist. Specifically for medical image segmentation a CNN with encoder-decoder architecture, named U-Net [9], was proposed and became a de facto standard solution for medical image semantic segmentation. There were several attempts to improve the U-Net architecture to boost the network performance for particular tasks [11,16]. However, an interesting observation was published in [6], claiming that the method configurations, such as pre-processing, have better potential to improve the network's performance than architectural modifications of the U-Net. The so-called not new U-Net (nnU-Net) dominated in 23 different segmentation challenges.

In this paper, we utilize the nnU-Net architecture and propose an improved version with residual short skip connections [2], and additional mixup augmentation [15], which was specifically tailored for the foot ulcer segmentation task. The fundamental goal was to demonstrate that this slight architectural modification can further boost overall performance, even for heavily imbalanced datasets with a wide variety of targeted structures.

The rest of the paper is organized as follows. In the next section, we provide a detailed data description. In the methodology section, the proposed solution is described. Finally, we present the results and discuss different aspects of our submission.

2   Dataset

The data used during our experiments were part of the official Diabetic Foot Ulcers Grand Challenge 2022 (DFUC2022) dataset gathered by the organizers [7,13]. They used three different cameras to capture the data. Kodak DX4530, Nikon D3300, and Nikon COOLPIX P100 were used to acquire close-ups of the whole foot at a distance of around 30–40 cm with the parallel orientation to the plane of an ulcer. Adequate room lights were used, instead of a camera flash, to get consistent colors in the images [12]. Additionally, to improve the performance and reduce computational costs of proposed models, the original size of the images, which varied between 1600 × 1200 and 3648 × 2736 pixels, was modified to 640 × 480 with the preservation of the aspect ratio. These images originated from the DFUC2020 dataset [1]. Moreover, photographs that were excessively out-of-focus and blurry were discarded to prevent misleading assumptions.

The ground-truth corresponding to the area of ulcer was produced by a podiatrist and a consultant physician specializing in the diabetic foot with more than five years of professional experience. In the case of disagreement, a third specialist podiatrist examined the photograph. Finally, the whole annotation process was completed using the polygonal shape tool within the VGG Image Annotator (VIA) application [3].

3   Methodology

Based on the previous successful applications of the nnU-Net architecture on various challenges, we chose its baseline version as the starting point for our experiments to segment ulcer structures. Authors of [2] introduced short skip connections as a solution for training very deep networks, which suffers from vanishing gradient problem and slow convergence. They also proved that these short skip connections could help the model to learn a better representation of targeted structures, thus improving overall performance. Based on these results, we decided to enhance the encoder part of the nnU-Net with this implementation of skip connections. Furthermore, we applied mixup augmentation on the input layer to further boost the overall performance. This technique was already shown to be beneficial for segmentation of 3D CT scans of abdominal area [4].

3.1   Preprocessing

The original DFUC2022 dataset consists of 2D photos. However, the standard nnU-Net pipeline can only work with 3D NIfTI data, so our first step was to transform the original images and masks to the required format. Here we applied two different approaches, one for the image data and one for masks. In the case of foot photos we added two additional dimensions, where the first one represented the color channel of the image and the second one was used to create a dummy 3D image with value one along the x-axis. Due to network requirements we additionally reordered the dimensions to be in the channel first format. The slightly different process was applied to masks. By default, the nnU-Net treats all pixels with value of 0 as background and everything with higher value as label. In order to fulfil this requirement we applied our custom thresholding method. Every pixel with a value below 255 was considered as background and the remaining pixels represented label.

Afterward, we performed standard preprocessing steps such as normalization, scaling, and other spatial transformations. In the case of the nnU-Net, these steps were automatically handled by nnU-Net processing. The image resampling strategy used third order spline interpolation for in-plane and nearest neighbors for out-of-plane. Global percentile clipping along with z-score with global foreground mean and standard deviation was chosen to normalize data. To clip HU values, the nnU-Net pipeline applied the default settings in the 0.5 to 99.5 percentile range.

3.2   Proposed Approach

We employ the U-Net architecture in all of our experiments. As the first modification, we implemented short skip connections in the encoder part to speed up the convergence process and boost the information flow through the individual intermediate layers. Primary, the encoder part of the nnU-Net is responsible for feature extraction and structure recognition. Thus the application of these additional skip connections in the decoder part is not necessary and would not contribute to the overall network performance.

As the second improvement, we propose utilization of the mixup augmentation method. According to the authors of [15], input images with ground truth labels denoted as (x_i, y_i), (x_j, y_j) are transformed to output tuples $(\hat{x}_i, \hat{y}_i)$ by the following transformations:

$$\hat{x} = \lambda x_i + (1 - \lambda)x_j \tag{1}$$

$$\hat{y} = \lambda y_i + (1 - \lambda)y_j \tag{2}$$

We hypothesize that this modification can prevent the issues related to imbalanced datasets, where the network has a tendency to segment the non-background portion of the image as background, thus ignoring targeted structures. We performed multiple experiments with regular mixup and its modified version named CutMix [14]. The key difference between these augmentations is that the regular mixup is applied on the whole mini-batch with a specified mixup ratio, while the CutMix version utilizes only the cropped part of the image, which is later mixed with another non-cropped image. The illustration of the mixup augmentations on original photos can be seen in Fig. 1.

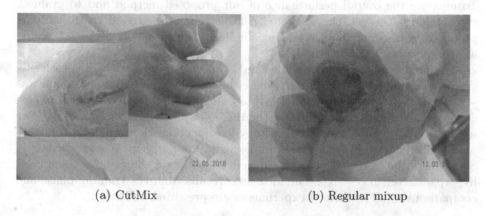

(a) CutMix   (b) Regular mixup

Fig. 1. Illustration of mixup augmentations on photographs from the DFUC2022 dataset.

Based on the obtained results, we decided to incorporate standard mixup augmentation with a mixup ratio of 50:50. This ratio was used for foot photos and also for labels. The mixup process for photos was straightforward, but not for the labels. Here we handled three different scenarios, based on the different meanings of pixel values. Pixels marked as background were only treated as background when they were mixed with background values from second label. In the remaining two scenarios the pixels after mixup were always considered as labels. So, in the end the weakest value the label could have after the mixing process was 127, therefore there was no risk of having weak masks that are more similar to the background than to the label and being marked as wound instead of background. The overview of the proposed method can be seen in Fig. 2.

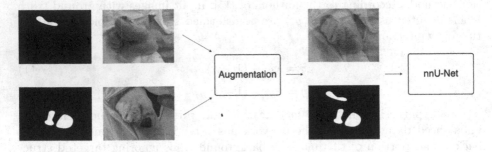

Fig. 2. Overview of proposed method

4   Experimental Results

To measure the overall performance of our proposed method and to evaluate the similarity between the predicted contours and the ground truth contours produced by the domain experts, Dice Similarity Coefficient (DSC) was selected as a primary statistical method.

The final version of our modified nnU-Net was trained with the stochastic gradient descent (SGD) optimizer, where the initial learning rate parameter was set to 0.01. We trained this model with a batch size of 2 for 1000 epochs, which is the default length of training for the nnU-Net pipeline. To minimize loss we used a combined Dice and Cross Entropy loss function. Here we did not further modify the short skip connections implementation and kept it as described in Sect. 3.2. In the case of the mixup augmentation, we set the value of the mixing hyperparameter to $\alpha = 0.5$. The measured results from the training phase and comparisons to all performed experiments are presented in Table 1 .

Table 1. Experimental results during the training phase on the DFU2022 public dataset

Network	Dice score
Baseline nnU-Net	0.6530
nnU-Net + mixup + skip connections (proposed solution)	**0.6713**
nnU-Net + skip connections	0.6565
nnU-Net + mixup	0.6616
nnU-Net + CutMix	0.6365

We also noticed that the individual implementation of short skip connections and mixup augmentation slightly enhanced segmentation quality. Based on this observation, we proposed a solution that is based on the combination of mixup and short skip connections. As a result, the proposed neural network significantly improved the overall performance of the nnU-Net and outperformed its baseline version. On the other hand, CutMix augmentation proved insufficient and performed even worse than baseline nnU-Net. We assume that this degradation was caused by extensive modification of the input data, which negatively impacted the learning process and thus resulted in misleading segmentation results.

To demonstrate the capability of our trained model, we randomly chose case 100630 from publicly available images. Here, we were capable of precisely segmenting ulcer structure with a Dice score of 0.9690. This example segmentation, depicted in Fig. 3, shows a diabetic foot ulcer in a common position. The picture was taken from above of the well-cleaned ulcer in good lighting conditions. The Dice score is high even though the ulcer was not healed at the time and despite the fact that the picture was taken before the necrectomy (removal of the dead tissue) was performed.

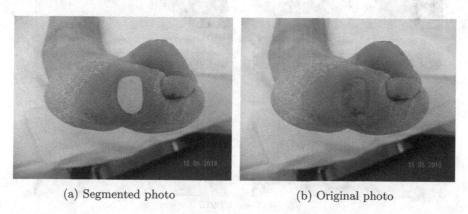

(a) Segmented photo   (b) Original photo

Fig. 3. Example segmentation for case 100630. Green color denotes the ground truth label, red denotes our segmentation result, and orange denotes the overlap of these two labels. (Color figure online)

Additionally, our model was evaluated on public test data during testing phase of the DFUC2022 challenge, where we achieved 9th place with a Dice score of 0.6975, which was not significantly different from 1st place with Dice score equal to 0.7287.

4.1 Fault-Case Analysis

In order to fully understand the capabilities of our proposed method, we have also performed fault-case analysis with a skilled clinician, who identified several sources of error in the segmentation results. Firstly, insufficient wound cleaning,

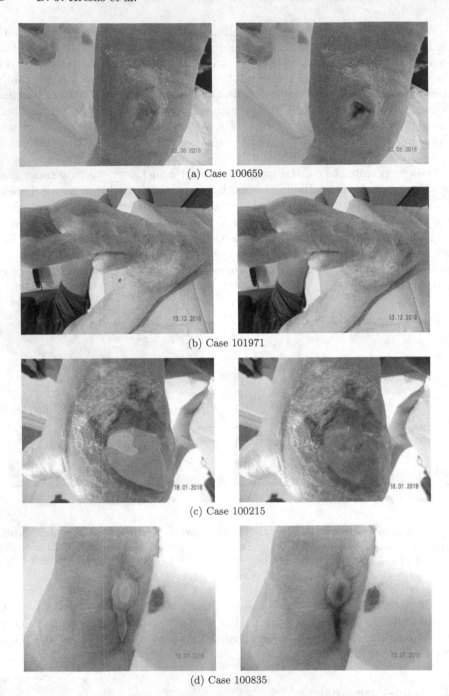

(a) Case 100659

(b) Case 101971

(c) Case 100215

(d) Case 100835

Fig. 4. Examples of faulty segmentation results. Segmented images are located on the left side and original images on the right side. Green color denotes ground truth label, red denotes our segmentation result, and orange denotes the overlap of these two labels. (Color figure online)

which led to capturing blood stains in the proximity of the ground truth wound as illustrated in Fig. 4a. Second, Fig. 4b illustrates atypical angles from which the pictures were taken. These resulted in wound morphology deformation. Third, an incomplete ground truth segmentation (some wounds are partially or entirely left out from the ground truth (Fig. 4c). On the other hand, there are also opposite examples, where the ground truth includes regions of healthy tissue as in Fig. 4d.

5   Conclusions

In this paper, we proposed an nnU-Net-based fully convolutional neural network for diabetic foot ulcer segmentation. The proposed approach minimizes the probability of over-fitting by using mixup augmentation. Additionally, we implemented a short skip connection into the nnU-Net architecture to improve training. The proposed solution achieved 9th place with a Dice segmentation score of 0.6975 on the test dataset of the DFU2022 challenge.

Acknowledgements. This work was supported by the Scientific Grant Agency of the Ministry of Education, Science, Research and Sport of the Slovak Republic and the Slovak Academy of Sciences under contract VEGA 1/0327/20.

References

1. Cassidy, B., et al.: The DFUC 2020 dataset: analysis towards diabetic foot ulcer detection. touchREVIEWS Endocrinol. **17**, 5–11 (2021). https://doi.org/10.17925/ EE.2021.17.1.5, https://www.touchendocrinology.com/diabetes/journal-articles/ the-dfuc-2020-dataset-analysis-towards-diabetic-foot-ulcer-detection/1
2. Drozdzal, M., Vorontsov, E., Chartrand, G., Kadoury, S., Pal, C.: The importance of skip connections in biomedical image segmentation. In: Carneiro, G., et al. (eds.) Deep Learning and Data Labeling for Medical Applications, pp. 179–187. Springer, Cham (2016)
3. Dutta, A., Zisserman, A.: The via annotation software for images, audio and video. In: Proceedings of the 27th ACM International Conference on Multimedia (2019)
4. Gazda, M., Bugata, P., Gazda, J., Hubacek, D., Hresko, D.J., Drotar, P.: Mixup augmentation for kidney and kidney tumor segmentation. In: Heller, N., Isensee, F., Trofimova, D., Tejpaul, R., Papanikolopoulos, N., Weight, C. (eds.) Kidney and Kidney Tumor Segmentation, pp. 90–97. Springer, Cham (2022)
5. Guo, J., Dardik, A., Fang, K., Huang, R., Gu, Y.: Meta-analysis on the treatment of diabetic foot ulcers with autologous stem cells. Stem Cell Res. Ther. **8**(1), 1–8 (2017)
6. Isensee, F., Jaeger, P.F., Kohl, S.A.A., Petersen, J., Maier-Hein, K.H.: NNU-net: a self-configuring method for deep learning-based biomedical image segmentation. Nat. Methods **18**(2), 203–211 (2021). https://doi.org/10.1038/s41592-020-01008-z, https://doi.org/10.1038/s41592-020-01008-z
7. Kendrick, C., et al.: Translating clinical delineation of diabetic foot ulcers into machine interpretable segmentation (2022). https://doi.org/10.48550/ARXIV. 2204.11618, https://arxiv.org/abs/2204.11618

8. Roche, M.M., Wang, P.P.: Factors associated with a diabetes diagnosis and late diabetes diagnosis for males and females. J. Clin. Transl. Endocrinol. **1**(3), 77–84 (2014)
9. Ronneberger, O., Fischer, P., Brox, T.: U-Net: convolutional networks for biomedical image segmentation. In: Navab, N., Hornegger, J., Wells, W.M., Frangi, A.F. (eds.) MICCAI 2015. LNCS, vol. 9351, pp. 234–241. Springer, Cham (2015). https://doi.org/10.1007/978-3-319-24574-4_28
10. Sun, H., et al.: IDF diabetes atlas: global, regional and country-level diabetes prevalence estimates for 2021 and projections for 2045. Diabetes Res. Clin. Pract. **183**, 109119 (2022)
11. Tsai, T.H., Huang, S.A.: Refined u-net: a new semantic technique on hand segmentation. Neurocomputing, **495**, 1–10 (2022). https://doi.org/10.1016/j.neucom.2022.04.079, https://www.sciencedirect.com/science/article/pii/S0925231222004696
12. Yap, M.H., Kendrick, C., Reeves, N.D., Goyal, M., Pappachan, J.M., Cassidy, B.: Development of diabetic foot ulcer datasets: an overview. Diabet. Foot Ulcers Grand Challenge **13183**, 1–18 (2021)
13. Yap, M.H., et al.: Diabetic foot ulcers grand challenge 2022 (2021). https://doi.org/10.5281/zenodo.6389665, https://doi.org/10.5281/zenodo.6389665
14. Yun, S., Han, D., Chun, S., Oh, S., Yoo, Y., Choe, J.: CutMix: regularization strategy to train strong classifiers with localizable features. In: 2019 IEEE/CVF International Conference on Computer Vision (ICCV), pp. 6022–6031. IEEE Computer Society, Los Alamitos, CA, USA (2019). https://doi.org/10.1109/ICCV.2019.00612, https://doi.ieeecomputersociety.org/10.1109/ICCV.2019.00612
15. Zhang, H., Cisse, M., Dauphin, Y.N., Lopez-Paz, D.: mixup: beyond empirical risk minimization (2017). https://doi.org/10.48550/ARXIV.1710.09412, https://arxiv.org/abs/1710.09412
16. Zhou, J., Lu, Y., Tao, S., Cheng, X., Huang, C.: E-RES U-Net: an improved u-net model for segmentation of muscle images. Expert Syst. Appl. **185**, 115625 (2021). https://doi.org/10.1016/j.eswa.2021.115625, https://www.sciencedirect.com/science/article/pii/S0957417421010198

DFU-Ens: End-to-End Diabetic Foot Ulcer Segmentation Framework with Vision Transformer Based Detection

Dariusz Kucharski[✉][ID], Aleksander Kostuch[ID], Filip Noworolnik, Andrzej Brodzicki[ID], and Joanna Jaworek-Korjakowska[ID]

Department of Automatic Control and Robotics, AGH UST, Krakow, Poland
{kucharski,kostuch,piekarski,brodzicki,jaworek}@agh.edu.pl
https://home.agh.edu.pl/~mdig/dokuwiki/doku.php

Abstract. Diabetic foot ulceration (DFU) is an open wound that occurs in approximately 15% of patients with diabetes, and is mostly located on the sole. Currently, it is one of the major challenges for healthcare systems around the world and a serious complication of diabetes. DFU causes infection and ischemia which can significantly prolong treatment and result in limb amputation and terminal illness. Thus, regular monitoring of the DFU area is necessary to assess the healing process and improve patient care. In this regard, we propose an end-to-end ensemble fully convolutional network (DFU-Ens), which mainly includes three modules: the U-Net module (DFU-Seg), the hybrid approach (DFU-Det) containing the YOLOv4 based detection module and the Vision Transformer DETR detection approaches. The ensemble solution is based on a high ranking strategy which combines DFU-Seg with a hybrid solution of bounding-box detection (performed using the latest DETR vision transformer architecture and YOLOv4) and patch segmentation. On the DFUC2022 validation set, we achieved 0.643 Dice score for the ensemble approach, 0.648 for DFU-Seg, and 0.556 and 0.581 for hybrid approaches based on YOLOv4 and DETR, respectively, with DETR having the best sensitivity. On the later published test set, DFU-Seg achieve a Dice score of 0.67 while the ensemble method achived 0.66.

Keywords: DFU · Diabetic foot · Segmentation · Detection · DETR · U-Net · YOLO

1 Introduction

Diabetes is a chronic disease that occurs either when the pancreas does not produce enough insulin or when the body cannot effectively use the insulin it produces. It is a very common disease which affects over 425 million people worldwide [7]. Based on World Health Organization data from 2019, diabetes was the direct cause of 1.5 million deaths and 48% of all deaths due to diabetes occurred before the age of 70. One of its most severe complications is the diabetic

© The Author(s), under exclusive license to Springer Nature Switzerland AG 2023
M. H. Yap et al. (Eds.): DFUC 2022, LNCS 13797, pp. 101–112, 2023.
https://doi.org/10.1007/978-3-031-26354-5_9

foot - a condition that can range from minor walking problems to a complete malfunction of parts of the nervous system resulting in an inability to feel pain or touch. The risk of a diabetic patient developing a foot ulcer within their lifetime is estimated to be 19–34% [2]. If not treated properly it often results in a limb amputation. Apart from balancing the diabetes itself, the treatment requires careful observation of the affected areas and very special care. This is especially important at early stages. Thus, there is a huge demand for CAD system dedicated for detection, precise segmentation as well as comparison and assessment of DFU areas. Such a system can be beneficial for several reasons: a) visual inspection can be subjective and inaccurate for DFU area segmentation, assessment and tissue classification, b) DFU examination observations are not always recorded in a consistent format and thus no comparison is possible, and c) manual assessment of DFU areas during the healing period is difficult without computer support.

Our contribution to this research area can be summarized as: (1) We propose an ensemble end-to-end DFU segmentation framework which includes three modules: the U-Net module (DFU-Seg), the hybrid approach (DFU-Det) containing the YOLOv4 based detection module, and the Vision Transformer DETR detection approaches; (2) We confirm that Vision Transformers can achieve high results for detection tasks in the medical domain; (3) we compare the outcomes of state-of-the-art models including an end-to-end U-Net architecture, detection with U-Net segmentation for YOLOv4 and DETR Vision Transformer for the DFU area segmentation task. We visualize the segmentation areas extracted by each architecture; and (4) Our solution achieved 0.643 Dice score for the DFU-Ens ensemble based approach, 0.648 for DFU-Seg, and 0.556 and 0.581 for the hybrid approaches based on YOLOv4 and DETR, respectively.

1.1   Related Work

Due to the contribution to this research area in a form of a Diabetic Foot Ulcer Challenge, which has been organised since 2020 [26], many research groups have addressed the DFU area assessment including detection (2020) [6] and classification (2021) [25]. Here, we present the most important research papers both regarding the DFU Challenge including classification and detection as well as general segmentation.

In DFUC2021 [5], the highest classification result of 0.6216 for F1-score and 0.8855 AUC has been achieved by [10]. The author also used Vision Transformers, but they were outperformed by Convolutional Neural Networks (CNN). Very interesting results and AI architectures with unique approaches have been presented by [1, 3, 13, 21]. In [3], the authors used an ensemble of EfficientNet architectures [24] and a pix2pix model [16] to generate artificial DFU wound images. In [21], the authors tested several pre-trained Vision Transformers and fine-tuned them for the classification task. It is, however, worth mentioning that to the best of our knowledge, there were no solutions participating in the challenge that used Vision Transformers for detection or segmentation.

For the segmentation task in general, there are many different solutions, ranging from classical computer vision techniques to sophisticated deep learning architectures including CNN and U-Net based models or Vision Transformers. The most popular architectures used for medical image segmentation are U-Nets [23] and autoencoders, with many different variations such as nnU-Net [15], CE-Net [12], UNET 3+ [14], or the latest Half-UNet [20]. The detection task, on the other hand is mostly dominated by RCNN-like methods, recently outperformed by faster solutions like YOLOv4-7 or SSD [19,22] as well as with the Vision Transformers [9]. Although not so widely used in the segmentation task itself, they already begin to outperform traditional convolutional networks in tasks such as classification and detection. In Sect. 2.3 we describe how this new solution, especially focusing on DETR architecture [4], might be utilised for DFU detection and segmentation.

2   DFU-Ens: End-to-End Ensemble DFU Segmentation Framework

The DFU segmentation task can vary from segmenting small areas to large ones that cover most of the entire image. With such a diverse dataset, it can be difficult to tune the network to perform well. Especially, there is a risk of false positive occurrences, as the network may focus on bigger, more definitive objects in the background. Therefore, as a solution to this size diversity, we propose the ensemble learning methodology. The ensemble learning methods are used to improve the results of deep learning by combining several models, which gives an improved prediction. Thus, the proposed DFU-Ens detection and segmentation framework uses an ensemble segmentation technique, which improves the accuracy of segmentation with enhanced performance.

The proposed DFU-Ens framework consists of three modules presented in Fig. 1: the end-to-end U-Net module (DFU-Seg), the hybrid approach (DFU-Det) containing the YOLOv4 based detection module, and the Vision Transformer DETR detection approach. Similar to the ensemble approach, the final prediction is the aggregation of the prediction made by each of the three proposed modules including the high ranking strategy based on combining DFU-Seg with a hybrid solution. This approach performs bounding-box detection using the latest DETR ViT architecture and YOLOv4 followed by patch segmentation. The iterative architecture adjustment, ensemble learning, and hyperparameter tuning have been used in this model, which has resulted in improvements to segmentation results.

2.1   Dataset: Diabetic Foot Ulcer Segmentation Challenge 2022

The Diabetic Foot Ulcer Segmentation Challenge 2022 [27] dataset consists of a training set of 2000 RGB images and a testing set of 2000 RGB images, where ulcer regions were delineated by experienced podiatrists and consultants. The DFUC2022 training set consists of 2304 ulcers, where the smallest ulcer size is

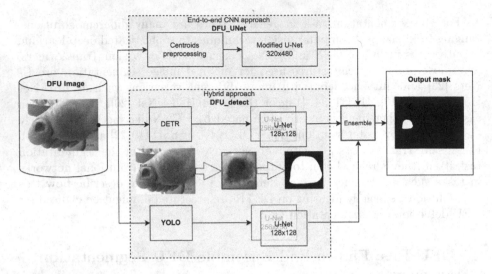

Fig. 1. The DFU-Ens architecture consists of three main modules: the U-Net part (DFU-Seg), the hybrid approach (DFU-Det) containing the YOLOv4 based detection module, and Vision Transformer DETR detection module. Before output each of them are combined using ensemble methods.

0.04% of the total image size, and the largest ulcer size is 35.04% of the total image size [17]. In the DFUC2022 dataset, 89% (2054 out of 2304) of the ulcers are less than 5% of the total image size. The smaller images in particular represent a significant challenge for segmentation algorithms as it is widely known that deep learning algorithms tend to omit small regions [11]. A small number of duplicates, which were annotated by different experts, were also added to the dataset [17]. More details on the DFUC organization, as well as live test leaderboard, can be found online at: https://dfu-2022.grand-challenge.org/.

2.2 Module I: End-to-End Segmentation Framework Based on the U-Net Architecture

The U-Net based architecture introduced in 2015 by [23] with many novel improvements is still today the most widely used in biomedical image segmentation tasks, often outperforming other solutions.

Data Preprocessing for Segmentation: As the models accept inputs smaller than the original image size found in the dataset, we take advantage and provide a diversified training set for each epoch, simply by extracting its subset. Firstly, we use the original mask in order to find the location of a reference object. Then, having its centroid and bounding box coordinates we randomly choose a patch around that centroid which serves as the input to the network. Randomness guarantees that for each epoch, for a particular image, an extracted example differs from a previous one from the same image to the previous epoch. Such an approach, together with typical augmentation methods like flips, re-scaling and translation, helped to reduce overfitting during training.

In order to provide the most representative examples during training, we divided patches extracted into three alternatives (Fig. 2). The first one randomly picks a new centroid which is inside the bounding box of the object. This guarantees that small objects should stay near the center of the patch, but never in the exact same place. The second one widens the interval where a new centroid is chosen from to the size of the patch. A result of this means that objects may appear on the edge of the patch. The last possibility randomly chooses a new centroid from the area of the whole image in order to provide examples with no objects on them. A new centroid is then used to extract a patch of neural network input size. During our experiments we achieved the best results for using steps with a ratio of 0.75:0.15:0.1 which means that in each batch there were, on average, 75% examples extracted using the first approach, 15% with the second, and 10% with the third.

After a single patch is extracted, it is converted to a CIELAB color space and then concatenated with the RGB representation. Finally, a 6-channel patch is extracted from an image constituting a training example to the network.

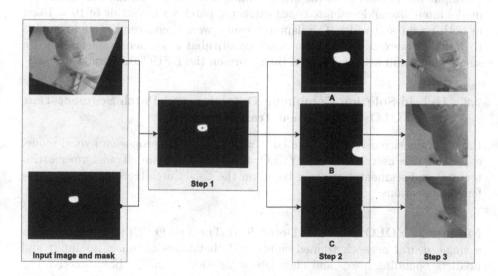

Fig. 2. Image preprocessing stage: Input image and mask augmentation, Step 1: centroid and bounding box coordinates of the object on the ground-truth mask, Step 2: randomly choose next step (2a, 2b or 2c) where 2a: centroid coordinates are drawn within the object's bounding box, 2b: alternative centroid is picked within range between the original centroid and patch size, 2c: new centroid randomly chosen on the whole image. Step 3: extracting the patch which constitutes an input to a network.

Training and Evaluation: In the end-to-end approach we took advantage of a U-Net and Xception hybrid architecture [8]. The input was defined as 320 *times* 480 px based on our experiments. Smaller network input, when compared to the original images, was used on purpose in order to use the preprocessing technique described above, mainly for the possibility of randomly choosing different patches from a single image.

For training, a 5-fold validation approach was used. The dataset was divided into 5 subsets, each of which consisted of 1600 examples for training and 400 for validation. Then five models were trained for 860 epochs, starting with $10e^{-3}$ learning rate and lowering it to $10e^{-5}$ after 130 epochs and then to $10e^{-5}$ after 650 epochs. As a cost function a combination of binary crossentropy and Jaccard Loss was proposed. Jaccard coefficient is defined as: $J(A,B) = |A \cup B|/|A \cap B|$, where A and B are reference set and model output set respectively. As Jaccard coefficient maximum value is 1.0 it can be used as a loss function as: $J_{loss} = 1 - J$. Because Jaccard works only for predictions which already have intersection with reference (for those without, it equals to 0 so there is no gradient to calculate weights update) a Binary Crossentropy loss (BCE_{loss}) was added. The final loss function was defined as: $Loss = 0.2 * BCE_{loss} + 0.8 * J_{loss}$.

For evaluation purposes, the DFUC2022 validation set was used. Because the input was smaller than the original images we used a sliding window of the model input size and evaluated each extracted patch with a stride of 16×16 px in both height and width. Overlapping results were then averaged and rounded to 1 if they were > 0.5. Such a result constituted a segmentation mask for a single model and achieved 0.6478 Dice score on the DFUC2022 validation set.

2.3 Hybrid Solution Combining Detection and Patch Segmentation with YOLOv4 and Vision Transformers

Based on the size diversity of the DFU wound areas we propose a hybrid model consisting of a detection stage (YOLOv4 and DETR Vision Transformer architecture) and segmentation stage based on the U-Net architecture, albeit with a few modifications.

Module 2: YOLOv4 Based Detection: The YOLO [22] approach involves a single neural network trained end-to-end that takes an image as input and predicts bounding boxes and class labels for each bounding box directly. The network was originally designed as a very fast solution, dedicated to real-time applications such as autonomous vehicles. Since then, it has become the most commonly used benchmark solution for almost all detection tasks. To improve and shorten the learning process the model was based on the YOLOv4 model pretrained on the COCO dataset [18]. Most importantly, instead of training one model for patch segmentation we tested with different patch sizes including 128 × 128 for smaller detections and 256 × 256 for larger ones. In Fig. 3 we present the outcomes for detection process and patch generation.

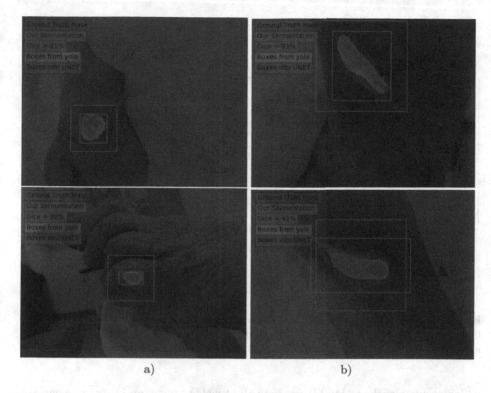

a)   b)

Fig. 3. Examples presenting the BBoxes obtained with YOLOv4 and segmentation results: a) small detections resized into patches of size 128 × 128, b) large detections resized into patches of size 256 × 256.

Module 3: DETR Vision Transformer Based Detection: DETR was introduced in 2020 by Facebook AI in the paper [4]. The framework consists of a CNN backbone network, a positional encoding module, an encoder-decoder architecture based on transformers, and prediction heads (Fig. 4). The pipeline utilizes bipartite matching between the predicted and ground-truth objects in order to drop post-processing of detections. Due to its parallel nature of processing predictions, DETR is very efficient and fast.

In order to increase accuracy, we decided to utilize a DETR model pretrained on the COCO dataset [18], which we fine-tuned on challenge data. We used the original loss with null class coefficient set to 0.05. We have left the default setting of $num\_queries = 100$. Even though the detection results obtained with DETR are fairly accurate, with our choice of parameters the number of false positives and duplicated detections is large. In order to drop unnecessary bounding boxes, instead of decreasing the number of queries, we consider only 3% of the highest confidence detections. This setting improves the detection of small ulcers and the ones barely distinguishable from the skin, and also enables filtering of the duplicate and false-positive predictions. In contrast to YOLO

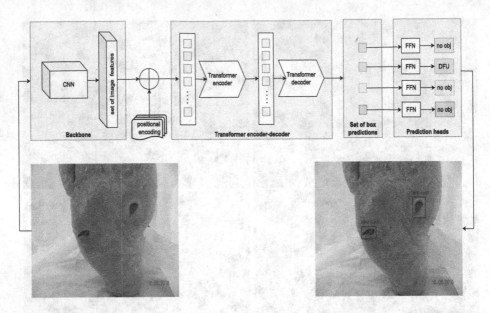

Fig. 4. DETR Architecture containing the most essential components: CNN ResNet-50 backbone, Transformer encoder-decoder and prediction heads with FFN (feed forward network), which is a 3-layer perceptron with ReLU activation function.

which requires anchors, DETR predicts all objects at once, and is trained end-to-end with a set loss function which performs bipartite matching between predicted and ground-truth objects. This can be a reason why the DETR model achieved higher precision, detecting images that were ignored by the YOLOv4 module (see Fig. 5).

Patch Adjustment and Segmentation: To address the problem of very different ulcer areas, we decided to use two sizes of patches: 128 × 128 for smaller objects and another of size 256 × 256 for larger ones. The segmentation architecture is based on U-Net as described in Sect. 2.2. Both models were trained using 5-fold cross-validation, and later verified on the DFUC2022 validation set. Training hyperparameters were optimised - the model was trained for 50 epochs, with Adam optimizer and a batch size of 8. Finally, the predictions of the 5-fold cross-validated model were thresholded to address the problem of under-segmentation. The confidence level was lowered to 0.2.

3   Experimental Results and Analysis

In this section we describe details of our experiments and present calculated statistics. Metrics such as Mean Overlap, Union Overlap, Dice Coefficient, Volume Similarity, False Negative Error, False Positive Error, and Jaccard Coefficient are calculated by uploading the mask results on the DFUC2022 Grand

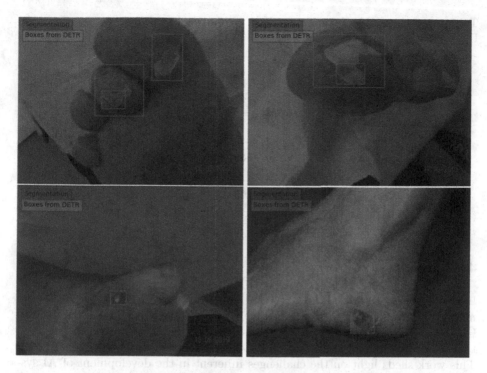

Fig. 5. Ulcer examples from the official DFUC2022 validation set that were missed by YOLOv4 and detected by DETR.

Challenge website, so they can be easily compared with other solutions taking part in the challenge.

In Table 1 we present our results for 3 modules: YOLOv4, DETR and DFU-Seg as well as the ensemble outcome for DFU-Ens. The DFU-Seg architecture achieved similar results to the ensemble method scoring 0.64 Dice score on the validation set. On the test set, those methods achieved a score of 0.67 and 0.66, respectively. We achieved 11th place on the leaderboard, with the winning solution achieving 0.72 Dice score.

However, we observed that the DFU-Seg model tended to omit some DFU areas while the DETR ViT architecture finds even the smallest changes in the skin, despite a very high margin of confidence threshold (0.97).

3.1 Hardware and Software

Models were trained on an NVIDIA TeslaV100 GPU, provided by PLGrid Infrastructure. Source code is publicly available at the following GitHub repository link:
https://github.com/thecookingmethods/dfuc22, accessed on 25.10.2022.

Table 1. Comparison of results obtained on the validation and test sets from the DFUC2022 Grand Challenge website.

Metric	Validations set				Test set	
	YOLOv4	DETR	DFU-Seg	DFU-Ens	DFU-Seg	DFU-Ens
MeanOverlap	0.5556	0.5808	**0.6479**	0.6432	**0.6725**	0.6566
UnionOverlap	0.4746	0.4790	**0.5483**	0.5471	**0.5690**	0.5555
DiceCoefficient	0.5556	0.5808	**0.6479**	0.6433	**0.6725**	0.6566
VolumeSimilarity	**0.2473**	0.0139	0.0898	0.1208	0.0290	**0.0889**
FalseNegativeError	0.3975	0.3622	0.2991	**0.2952**	0.2830	**0.2828**
FalsePositiveError	0.3931	0.3293	**0.2945**	0.3020	**0.2555**	0.2818
JaccardCoefficient	0.4746	0.4790	**0.5483**	0.5471	**0.5690**	0.5555

4   Conclusion

In this paper, we proposed the DFU-Ens architecture which consists of three segmentation approaches which can effectively detect and extract DFU wound areas. On the DFUC2022 test set, our methods achieved 0.67 Dice score. The segmentation results highlight the difficulty in DFU clinical delineation and assessment. This work sheds light on the challenges inherent in the development of AI systems which can aid the standardisation of DFU delineation over time to track healing progress. In future work we will analyze in-depth the obtained results, propose data augmentation for omitted cases, and explore deep semi-supervised learning methods to enhance detection and segmentation accuracy.

Acknowledgements. Research project partly supported by program "Excellence initiative - research university" for the AGH University of Science and Technology and by PLGrid Infrastructure.

References

1. Ahmed, S., Naveed, H.: Bias adjustable activation network for imbalanced data—diabetic foot ulcer challenge 2021. In: Yap, M.H., Cassidy, B., Kendrick, C. (eds.) DFUC 2021. LNCS, vol. 13183, pp. 50–61. Springer, Cham (2022). https://doi.org/10.1007/978-3-030-94907-5_4
2. Armstrong, D.G., Boulton, A.J.M., Bus, S.A.: Diabetic foot ulcers and their recurrence. N. Engl. J. Med. **376**(24), 2367–2375 (2017)
3. Bloch, L., Brüngel, R., Friedrich, C.M.: Boosting efficientnets ensemble performance via pseudo-labels and synthetic images by pix2pixHD for infection and Ischaemia classification in diabetic foot ulcers. In: Yap, M.H., Cassidy, B., Kendrick, C. (eds.) DFUC 2021. LNCS, vol. 13183, pp. 30–49. Springer, Cham (2022). https://doi.org/10.1007/978-3-030-94907-5_3
4. Carion, N., Massa, F., Synnaeve, G., Usunier, N., Kirillov, A., Zagoruyko, S.: End-to-end object detection with transformers (2020). ArXiv: abs/2005.12872

5. Cassidy, B., Kendrick, C., Reeves, N.D., Pappachan, J.M., O'Shea, C., Armstrong, D.G., Yap, M.H.: Diabetic foot ulcer grand challenge 2021: evaluation and summary. In: Yap, M.H., Cassidy, B., Kendrick, C. (eds.) DFUC 2021. LNCS, vol. 13183, pp. 90–105. Springer, Cham (2022). https://doi.org/10.1007/978-3-030-94907-5_7

6. Cassidy, B., et al.: The DFUC 2020 dataset: analysis towards diabetic foot ulcer detection. TouchREVIEWS in Endocrinol. **17**(1), 5–11 (2021)

7. Cho, N.H., et al.: IDF diabetes atlas: global estimates of diabetes prevalence for 2017 and projections for 2045. Diabetes Res. Clin. Pract. **138**, 271–281 (2018)

8. Chollet, F., et al.: Keras (2015). https://github.com/fchollet/keras

9. Dosovitskiy, A., et al.: An image is worth 16x16 words: transformers for image recognition at scale (2021). ArXiv: abs/2010.11929

10. Galdran, A., Carneiro, G., Ballester, M.A.G.: Convolutional nets versus vision transformers for diabetic foot ulcer classification. In: Yap, M.H., Cassidy, B., Kendrick, C. (eds.) DFUC 2021. LNCS, vol. 13183, pp. 21–29. Springer, Cham (2022). https://doi.org/10.1007/978-3-030-94907-5_2

11. Goyal, M., Yap, M.H.: Multi-class semantic segmentation of skin lesions via fully convolutional networks (2020). ArXiv: abs/1711.10449

12. Gu, Z., et al.: Ce-Net: context encoder network for 2D medical image segmentation. IEEE Trans. Med. Imaging **38**, 2281–2292 (2019)

13. Güley, O., Pati, S., Bakas, S.: Classification of infection and ischemia in diabetic foot ulcers using VGG architectures. In: Yap, M.H., Cassidy, B., Kendrick, C. (eds.) DFUC 2021. LNCS, vol. 13183, pp. 76–89. Springer, Cham (2022). https://doi.org/10.1007/978-3-030-94907-5_6

14. Huang, H., et al.: Unet 3+: a full-scale connected Unet for medical image segmentation. In: ICASSP 2020–2020 IEEE International Conference on Acoustics, Speech and Signal Processing (ICASSP), pp. 1055–1059 (2020)

15. Isensee, F., Jaeger, P.F., Kohl, S.A.A., Petersen, J., Maier-Hein, K.: nnU-Net: a self-configuring method for deep learning-based biomedical image segmentation. Nat. Methods **18**, 203–211 (2020)

16. Isola, P., Zhu, J.Y., Zhou, T., Efros, A.A.: Image-to-image translation with conditional adversarial networks. In: 2017 IEEE Conference on Computer Vision and Pattern Recognition (CVPR), pp. 5967–5976 (2017)

17. Kendrick, C., et al.: Translating clinical delineation of diabetic foot ulcers into machine interpretable segmentation (2022). ArXiv: abs/2204.11618

18. Lin, T.Y., et al.: Microsoft COCO: common objects in context. In: Fleet, D., Pajdla, T., Schiele, B., Tuytelaars, T. (eds.) ECCV 2014. LNCS, vol. 8693, pp. 740–755. Springer, Cham (2014). https://doi.org/10.1007/978-3-319-10602-1_48

19. Liu, W., et al.: SSD: single shot multibox detector. In: Leibe, B., Matas, J., Sebe, N., Welling, M. (eds.) ECCV 2016. LNCS, vol. 9905, pp. 21–37. Springer, Cham (2016). https://doi.org/10.1007/978-3-319-46448-0_2

20. Lu, H., She, Y., Tie, J., Xu, S.: Half-UNet: a simplified u-net architecture for medical image segmentation. Front. Neuroinf. **16** (2022)

21. Qayyum, A., Benzinou, A., Mazher, M., Meriaudeau, F.: Efficient multi-model vision transformer based on feature fusion for classification of DFUC2021 challenge. In: Yap, M.H., Cassidy, B., Kendrick, C. (eds.) DFUC 2021. LNCS, vol. 13183, pp. 62–75. Springer, Cham (2022). https://doi.org/10.1007/978-3-030-94907-5_5

22. Redmon, J., Divvala, S.K., Girshick, R.B., Farhadi, A.: You only look once: unified, real-time object detection. In: 2016 IEEE Conference on Computer Vision and Pattern Recognition (CVPR), pp. 779–788 (2016)

23. Ronneberger, O., Fischer, P., Brox, T.: U-Net: Convolutional networks for biomedical image segmentation (2015). ArXiv: abs/1505.04597
24. Tan, M., Le, Q.V.: EfficientNet: rethinking model scaling for convolutional neural networks (2019). ArXiv: abs/1905.11946
25. Yap, M.H., Cassidy, B., Pappachan, J.M., O'Shea, C., Gillespie, D., Reeves, N.D.: Analysis towards classification of infection and ischaemia of diabetic foot ulcers. In: 2021 IEEE EMBS International Conference on Biomedical and Health Informatics (BHI), pp. 1–4 (2021)
26. Yap, M.H., et al.: Diabetic foot ulcers grand challenge 2020 (2020). https://doi.org/10.5281/zenodo 3715020
27. Yap, M.H., et al.: Diabetic foot ulcers grand challenge 2022, March 2021. https://doi.org/10.5281/zenodo.6389665

Summary Paper

Diabetic Foot Ulcer Grand Challenge 2022 Summary

Connah Kendrick[1](✉)(iD), Bill Cassidy[1](iD), Neil D. Reeves[2](iD),
Joseph M. Pappachan[3](iD), Claire O'Shea[4], Vishnu Chandrabalan[3](iD),
and Moi Hoon Yap[1](iD)

[1] Department of Computing and Mathematics, Manchester Metropolitan University,
Manchester M1 5GD, UK
`Connah.Kendrick@mmu.ac.uk`
[2] Musculoskeletal Science and Sports Medicine, Manchester Metropolitan University,
Manchester M1 5GD, UK
[3] Lancashire Teaching Hospitals NHS Foundation Trust, Preston PR2 9HT, UK
[4] Waikato District Health Board, Hamilton 3240, New Zealand

Abstract. The Diabetic Foot Ulcer Challenge 2022 focused on the task
of diabetic foot ulcer segmentation, based on the work completed in
previous DFU challenges. The challenge provided 4000 images of full-
view foot ulcer images together with corresponding delineation of ulcer
regions. This paper provides an overview of the challenge, a summary of
the methods proposed by the challenge participants, the results obtained
from each technique, and a comparison of the challenge results. The best-
performing network was a modified HarDNet-MSEG, with a Dice score
of 0.7287.

1 Introduction

Initial works [1,2] in diabetic foot ulcer (DFU) analysis demonstrated that deep
neural networks had the potential to give reliable classification and segmentation
outcomes for DFU regions. However, these works were limited by the small
dataset size (705 images), low image resolution (500 × 500 pixels) and non-
clinical annotations. In more recent works, Yap et al. [3,4] hosted the first DFU
detection and classification challenges, in conjunction with MICCAI 2020 and
MICCAI 2021. Then, followed by Wang et al. [5,6] who hosted an online MICCAI
2021 event in The Foot Ulcer Segmentation Challenge. They combined a series
of datasets from the AZH wound care centre and developed a 2020 dataset
containing 1109 224 × 224 images. Then, they released a newer dataset [6] for
the FUSeg challenge competition, containing 1210 images, with a resolution of
512 × 512 pixels. In addition, 160 images from the Medetec wound dataset was
included, resized to 560 × 391 pixels. The data was annotated by clinical staff at
the AZH foot clinic, but was low resolution and included black borders to pad
the images.

In 2022, Kendrick et al. [7] introduced the DFUC2022 dataset, which contains 2000 high resolution DFU images with clinical annotation, and 2000 images

M. H. Yap et al. (Eds.): DFUC 2022, LNCS 13797, pp. 115–123, 2023.
https://doi.org/10.1007/978-3-031-26354-5_10

without annotation for use as the test set. When compared to last year's challenge [4], the total number of participating countries has increased from 25 to 47. Figure 1 highlights the distribution of the DFUC2022 dataset users.

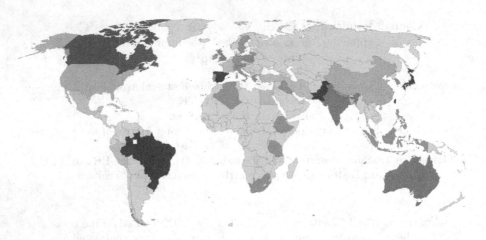

Fig. 1. Distribution of researchers by country of origin who used the DFUC2022 datasets.

2   Methodology

This section summarises the creation of the DFUC2022 dataset, the performance metrics, and analysis of the methods proposed by the participants of the challenge.

2.1   Datasets and Ground Truth

The DFUC2022 dataset is the largest publicly available DFU segmentation dataset and has the highest resolution images. The dataset consists of 4000 images collected from diabetic foot clinics at the Lancashire Teaching Hospitals NHS Foundation Trust, UK. Each image was collected using one of three cameras by medical photographers during patient appointments. During each appointment, medical photographers took photographs of the ulcers at a distance of approximately 30–40 cm without the aid of a tripod [8,9]. In some cases, the photograph shows a consistent background (blue or white), but others showed no background. These issues highlight the challenges inherent in manual medical photography in clinical settings.

The dataset was annotated by 5 podiatrists using the VGG annotator [10,11] application to create polygon outlines of the DFU regions. The dataset was then processed using an active contour algorithm [12] which smoothed the manually delineated regions that were drawn around the DFU regions. For further details, please refer to Kendrick et al. [7].

For DFUC2022, we split the dataset into two sets of 2000 images, one for training, and the second for validation and testing. The training set comprises a variety of DFU cases, where individual feet may exhibit multiple DFUs, resulting in a total of 2304 ulcers in the training set. The dataset comprised cases of the same wound at different stages of development, whereby cases may be healing or may have becoming more severe. Due to the whole foot nature of the images and some early cases, a large portion (89%) of the DFU wounds were less than 5% of the total image size. However, the dataset still comprised DFU within the size range between 0.04%–35.04% of the total image size.

2.2 Performance Metrics

We compared the performance of the deep learning networks on the accuracy of the segmentation using Dice coefficient, Jaccard, False Positive Error (FPE) and False Negative Error (FNE). We used image-based metrics so that multiple lesions in a single image were treated as one whole lesion. To determine the leaderboard ranking we used the Dice coefficient 1. The equation is based on Intersection over Union (IoU), but doubles the amount of pixels. This provides the metric with preserved sensitivity and resistance to outliers. In addition to providing a value between 0–1, where 0 is no overlap and 1 is an exact overlap.

$$Dice = 2 * \frac{|X \cap Y|}{|X| + |Y| - |X \cap Y|} \tag{1}$$

We also included the Jaccard index, also called IoU. The metric compares the area of overlap over the area of union. It also outputs values similar to Dice.

$$IoU = \frac{|X \cap Y|}{|X| + |Y|} \tag{2}$$

The FPE indicates the percentage in which a selected system falsely predicts a none DFU pixel as DFU.

$$FPE = \frac{FP}{FP + TN} \tag{3}$$

The FPE indicates the percentage in which a selected system falsely predicts a DFU pixel as none DFU.

$$FNE = \frac{FN}{FN + TP} \tag{4}$$

For both the Dice and the Jaccard metrics, the metrics assume overlap has occurred, if no overlap is present we score 0 for that image.

2.3   Summary of the Proposed Methods

In this section, we summarise the methods used by the top 10 entries for the DFUC2022. We present the methods in reverse order, i.e., start from the 10th position and through the leaderboard top rank. We provide a brief overview of the methods used and the scores that each achieved.

The method that ranked 10th in the challenge was submitted by AGH_MVG (the team from AGH University of Science and Technology, Poland). They used an end-to-end ensemble of 3 models in their proposed pipeline, namely, YOLOv4 [13] to localise the DFU, and a Vision Transformer DETR [14] with a U-Net module to segment the detected DFU regions. They used an augmentation technique to train the model in which the extracted DFU region has different percentages of visibility allowing the network to learn from semi-occluded images. Their approach achieved a Dice score of 0.6725.

The method which placed ninth was submitted by IISLab (Technical University of Kosice, Kosice, Slovakia and Department of Burns and Reconstructive Surgery, PJ Safarik University and Hospital Agel, Kosice-Saca, Slovakia). They modified the nnU-Net [15] architecture to work with 2D data rather than 3D data that the model had originally been designed for. They implemented two versions of the model, with and without skip connections to bridge data between convolutional and deconvolutional layers of the U-Net. Additionally, they analysed the use of cutmix [16] and mixup [17] augmentations and observed that using a combination of both increased the performance of the network. Overall, their best model achieved a Dice score of 0.6975.

The eighth place method was submitted by DGUT-XP(Dongguan University Of Technology, China). They achieved a Dice score of 0.6894. However, they did not submit their method description for inclusion into the summary paper.

The method in seventh place was submitted by GP_2022 (Alexandria University, Egypt and University of Southern California, USA). They used an ensemble of two models: DeepLabV3+ [18] and SegFormer [19]. They used pretraining for both models with the ade20k dataset [20], but with different optimisers. They applied standard augmentation techniques to the images during training, and ensembled the results from both networks to produce the final prediction. Additionally, they included a post-processing stage for hole filling and the removal of small pixel regions. They achieved a Dice score of 0.6986.

The method in sixth place was submitted by FDHO (University of Applied Sciences and Arts Dortmund, Germany). They used a combined-refinement approach where they generated realistic DFU images using StyleGan+ADA [21]. They also trained a series of models based on a Feature Pyramid Network using the modified ResNeXt-101 backbone to include the squeeze and excite layer. The models were used to annotate the generated images, which were used to train a series of extended models under 5-fold cross validation. To aid the process, they performed data augmentation over the dataset. In addition, they used post-processing to remove small pixels regions in the prediction results. They achieved a FID score of 19.09 on the generated images, demonstrating the ability to generate a diverse dataset as well as a Dice score of 0.7169 for the ensemble of extended models.

The method which placed fifth was submitted by Seoyoung (Sogang University, South Korea). They used a Swim Transformer based backbone for the upper network. They padded the dataset by using a series of augmentation techniques resulting in a total of 8000 training images which included the original training images. They also experimented using DeepLabV3+, OCRNet, and ConvNeXt [22]. Using their approach, they achieved a Dice score of 0.7220, however, the author did not submit a full paper.

The method in fourth place was submitted by Adar-Lab (National Yang Ming Chiao Tung University, Taiwan). They implemented a two-stage approach, firstly they used Fast-RCNN [23] to perform object detection on the high-resolution image. From the Fast-RCNN model they extract regions of interest, then they used TransFuse [24] for segmentation. In their experiment, they use different fusion techniques to merge transformer and CNN outputs. They used a unique augmentation approach for the model architecture, where for every original image, they cropped 4 images from each corner, with the size of 384×384 pixels. In their training, they used a combined loss of pixel-based Binary Cross-Entropy (BCE) and IoU. They achieved a Dice score of 0.7254.

The method placed third was submitted by agaldran (Universitat Pompeu Fabra, Spain). They implemented a two-length cascade of encoder decoder networks, the encoders were pretrained on imagenet, whereas the decoders used feature-pyramid networks [25], which improves the feature extraction of DFU at different sizes. Their experiments showed that the ResNeXt101 model had the best performance. They further increased the performance by performing a study of loss metrics, i.e., BCE training only, Dice training only, BCE with slow interpolation to Dice, BCE and a hard switch to Dice, and BCE+Dice training. They showed that for DFU and polyp segmentation the combined BCE+Dice work performed best and provided valuable insights into the effect of these combined losses on the training process. They achieved a Dice score of 0.7263.

The method which placed second was submitted by LkRobotAILab (West China Hospital, China). Their method is based on the state-of-the-art segmentation network OCRNet [26]. They completed an ablation study using an alternative model backbone that uses a mix of convolutional and transformer-based structures. Their study showed improvements to network performance when using ConvNeXt-XLarge with pre-training on ImageNet-21K. The dataset images were processed using gamma correction to help standardise during inference. Finally, to refine network performance, they used boundary loss which focuses on the edges of the predicted segmentation regions. They achieved a Dice score of 0.7280.

The winning entry for DFUC2022 was submitted by yllab (National Tsing Hua University, Taiwan). Their method was inspired by the state-of-the-art in polyp segmentation in colonoscopy videos. The network was based on HarDNet-MSEG [27], which uses an encoder-decoder structure. The encoder improved upon the Densenet [28] structure by reducing the number of shortcuts and increasing channel count. The decoder takes input from the encode at multiple stages through specialised Receptive Field Block (RFB) modules which are then aggregated via dense layers and upsampled for the final segmentation result. They improved the HarDNet-MSEG model by balancing the inputs and outputs of the encoder by splitting the channels of the input, as well as adjusting the skip connections to different layers. The decoder is then replaced with a Lawin Transformer [29]. They then use post processing to fill holes in the resulting segmentation map. Using this method, they achieved the best Dice score of 0.7287.

3   Results and Discussion

In total, we received 1320 submissions throughout the whole challenge process. In addition, a total of 22 teams participated in the challenge, with 50 individual users registered on the DFUC2022 Grand Challenge website. There were approximately 1100 submissions for the validation stage. During the validation stage, we allowed a maximum of 10 submissions per-day with a maximum of 200 prediction masks per submission. The best performing team for the validation stage was LkRobotAILab, with the best Dice score of 0.7156, a Jaccard score of 0.6195, and a FNE score of 0.2352. It is noted that team seoyoung achieved the best FPE score of 0.2312.

During the release of the test set, we observed that the positions on the leaderboard changed. Table 1 compares the performance of the Top-10 submissions. Yllab was the winning team, with a Dice score of 0.7287, an improvement of 0.1579 over the baseline scores of the test set (0.5708), provided by the organisers [7]. LkRobotAILab scored the best Jaccard score of 0.6276, with 0.1727 improvement over the baseline scores. Agadran scored 0.2210 on FPE, with 0.1614 improvement over the baseline. Finally, Adar-Lab scored 0.1847 for FNE, with 0.0654 improvement over the baseline result.

The results demonstrate an improvement over the baseline scores in most cases. This highlights the ability of the proposed techniques to segment DFU regions. However, in the case of FNE, we only observed a minor improvement on the scores. This is most likely due to the ratio of DFU pixels in the images, compared to the none-DFU pixels. As highlighted by some participants, duplication identification, where the same image had been labelled by more than one clinician, demonstrated disagreement between annotators which could potentially improve the performance of segmentation if resolved. After the challenge, the organisers have opened a live leaderboard, where at the time of writing this paper, the best performance on live leaderboard is reported by Kendrick et al. [7], with Dice Score of 0.7446.

Table 1. The top-10 participating teams in DFUC2022, starting with the best Dice score. † = higher score is better; ⊎ = lower score is better. **Bold** indicates the best overall result.

Team	Metrics			
	Dice †	Jaccard †	FPE ⊎	FNE ⊎
yllab	**0.7287**	0.6252	0.2048	0.2341
LkRobotAILab	0.7280	**0.6276**	0.2154	0.2261
Agadran	0.7263	0.6273	0.2262	**0.2210**
Adar-Lab	0.7254	0.6245	**0.1847**	0.2582
seoyoung	0.7220	0.6208	0.1925	0.2584
FHDO	0.7169	0.6130	0.214	0.2453
GP_2022	0.6986	0.5921	0.2065	0.2778
DGUT-XP	0.6984	0.5945	0.2523	0.2379
IISlab	0.6974	0.5926	0.2163	0.2734
AGH_MVG	0.6725	0.5690	0.2555	0.2830

4   Conclusion

In this study, we introduced the largest publicly available DFU segmentation dataset, annotated by clinical experts working in diabetic foot clinics. We propose a supervised technique to track the wound progression and recovery at point of care. This dataset is not the first publicly available DFU segmentation dataset, however, our dataset contains the largest number of images at a higher resolution than those datasets previously released. We provide this new dataset to the research community to encourage further research and progress in the field. The advancements shown in our proposed method has the potential to be used to support clinical staff in hospital settings and to assist with at-home tracking of DFU by patients. Thus, aiding the treatment process and preventing future complications such as infection and resulting limb amputation.

The networks demonstrated in this paper show that although deep learning can provide high level results, there is still much work to do to in the task of automated segmentation of DFU wounds. These challenges include the ability to segment early onset and small DFU regions in whole foot images, the differential between DFU and peri-wound regions, and contextual wound information such as signs of healing and degradation.

This work can help to progress the field by using such models as part of fully automated DFU monitoring systems that can be used at home or in foot clinics to track DFU healing progress and to relieve burden on healthcare workers. This work will build on our existing framework [30] in delivering an easy-to-use system capable of advanced forms of diabetic foot analysis, which will include longitudinal monitoring as a means of assessing wound healing progress.

Acknowledgment. We would like to thank the MICCAI conference for hosting DFUC2022, and AITIS for sponsoring the wining teams' prizes. We would also like to thank all participants of the challenge for their effort and contributions to DFU research.

References

1. Goyal, M., Reeves, N.D., Rajbhandari, S., Ahmad, N., Wang, C., Yap, M.H.: Recognition of ischaemia and infection in diabetic foot ulcers: dataset and techniques. Comput. Biol. Med. **117**, 103616 (2020)
2. Goyal, M., Yap, M. H., Reeves, N. D., Rajbhandari, S., Spragg, J.: Fully convolutional networks for diabetic foot ulcer segmentation. In: 2017 IEEE International Conference on Systems, Man, and Cybernetics (SMC), pp. 618–623. IEEE (2017)
3. Yap, M.H., et al.: Deep learning in diabetic foot ulcers detection: a comprehensive evaluation. Comput. Biol. Med. **135**, 104596 (2021)
4. Yap, M.H., et al.: Diabetic foot ulcers grand challenge, p. 3715020 (2020). https://doi.org/10.5281/zenodo
5. Wang, C., et al.: Fully automatic wound segmentation with deep convolutional neural networks. Sci. Rep. **10**(1), 1–9 (2020)
6. Rostami, B., Anisuzzaman, D.M., Wang, C., Gopalakrishnan, S., Niezgoda, J., Yu, Z.: Multiclass wound image classification using an ensemble deep CNN-based classifier. Comput. Biol. Med. **134**, 104536 (2021)
7. Kendrick, C., et al.: Translating clinical delineation of diabetic foot ulcers into machine interpretable segmentation. arXiv preprint arXiv:2204.11618 (2022)
8. Yap, M.H., Kendrick, C., Reeves, N.D., Goyal, M., Pappachan, J.M., Cassidy, B.: Development of diabetic foot ulcer datasets: an overview. In: Yap, M.H., Cassidy, B., Kendrick, C. (eds.) DFUC 2021. LNCS, vol. 13183, pp. 1–18. Springer, Cham (2022). https://doi.org/10.1007/978-3-030-94907-5_1
9. Pappachan, J.M., Cassidy, B., Fernandez, C.J., Chandrabalan, V., Yap, M.H.: The role of artificial intelligence technology in the care of diabetic foot ulcers: the past, the present, and the future. World J. Diab. **13**(12), 1131–1139 (2022)
10. Dutta, A., Gupta, A., Zissermann, A.: VGG image annotator (via) 2016. http://www.robots.ox.ac.uk/vgg/software/via/
11. Dutta, A., Zisserman, A.: The VIA annotation software for images, audio and video. In: Proceedings of the 27th ACM International Conference on Multimedia, MM 2019, New York, NY, USA, ACM (2019)
12. Kroon, D.-J.: Snake: active contour (2022)
13. Shinde, S., Kothari, A., Gupta, V.: Yolo based human action recognition and localization. Procedia Comput. Sci. **133**, 831–838 (2018)
14. Xu, W., Xu, Y., Chang, T., Tu, Z.: Co-scale conv-attentional image transformers. In: Proceedings of the IEEE/CVF International Conference on Computer Vision, pp. 9981–9990 (2021)
15. Isensee, F., et al.: Abstract: nnU-Net: self-adapting framework for U-Net-based medical image segmentation. In: Bildverarbeitung für die Medizin 2019. I, pp. 22–22. Springer, Wiesbaden (2019). https://doi.org/10.1007/978-3-658-25326-4_7
16. Yun, S., Han, D., Oh, S. J., Chun, S., Choe, J., Yoo, Y.: Cutmix: regularization strategy to train strong classifiers with localizable features. In: Proceedings of the IEEE/CVF international conference on computer vision, pp. 6023–6032 (2019)
17. Zhang, H., Cisse, M., Dauphin, Y.N., Lopez-Paz, D.: Mixup: beyond empirical risk minimization. In: International Conference on Learning Representations (2018)

18. Chen, L.C., Papandreou, G., Kokkinos, I., Murphy, K., Yuille, A.L.: DeepLab: semantic image segmentation with deep convolutional nets, Atrous convolution, and fully connected CRFs. IEEE Trans. Pattern Anal. Mach. Intell. **40**(4), 834–848 (2018)
19. Xie, E., Wang, W., Yu, Z., Anandkumar, A., Alvarez, J.M., Luo, P.: SegFormer: simple and efficient design for semantic segmentation with transformers. In: Advances in Neural Information Processing Systems, vol. 34, pp. 12077–12090 (2021)
20. Zhou, B., Zhao, H., Puig, X., Fidler, S., Barriuso, A., Torralba, A.: Scene parsing through ade20k dataset. In: Proceedings of the IEEE Conference on Computer Vision and Pattern Recognition, pp. 633–641 (2017)
21. Karras, T., Aittala, M., Hellsten, J., Laine, S., Lehtinen, J., Aila, T.: Training generative adversarial networks with limited data. In: Advances in Neural Information Processing Systems, vol. 33, pp. 12104–12114 (2020)
22. Liu, Z., Mao, H., Wu, C.Y., Feichtenhofer, C., Darrell, T., Xie, S.: A convnet for the 2020s. In: Proceedings of the IEEE/CVF Conference on Computer Vision and Pattern Recognition, pp. 11976–11986 (2022)
23. Girshick, R.: Fast r-CNN. In: Proceedings of the IEEE International Conference on Computer Vision, pp. 1440–1448 (2015)
24. Zhang, Yundong, Liu, Huiye, Hu, Qiang: TransFuse: fusing transformers and CNNs for medical image segmentation. In: De Bruijne, M., et al. (eds.) MICCAI 2021. LNCS, vol. 12901, pp. 14–24. Springer, Cham (2021). https://doi.org/10.1007/978-3-030-87193-2_2
25. Lin, T.Y., Dollár, P., Girshick, R., He, K., Hariharan, B., Belongie, S.: Feature pyramid networks for object detection. In: Proceedings of the IEEE Conference on Computer Vision and Pattern Recognition, pp. 2117–2125 (2017)
26. Liu, H., Liu, F., Fan, X., Huang, D.: Polarized self-attention: towards high-quality pixel-wise regression. arXiv preprint arXiv:2107.00782 (2021)
27. Huang, C.H., Wu, H.Y., Lin, Y.L.: HarDNet-MSEG: a simple encoder-decoder polyp segmentation neural network that achieves over 0.9 mean dice and 86 FPS, pp. 1–13 (2021)
28. Huang, G., Liu, Z., Van Der Maaten, L., Weinberger, K.Q.: Densely connected convolutional networks. In: Proceedings of the IEEE Conference on Computer Vision and Pattern Recognition, pp. 4700–4708 (2017)
29. Yan, H., Zhang, C., Wu, M.: Lawin transformer: improving semantic segmentation transformer with multi-scale representations via large window attention. arXiv preprint arXiv:2201.01615 (2022)
30. Cassidy, B., et al.: A cloud-based deep learning framework for remote detection of diabetic foot ulcers. IEEE Pervasive Comput. **21**, 78–86 (2022)

Author Index

M. H. Yap et al. (Eds.): DFUC 2022, LNCS 13797, p. 125, 2023.
https://doi.org/10.1007/978-3-031-26354-5

Printed in the United States
by Baker & Taylor Publisher Services

Printed in the United States
by Baker & Taylor Publisher Services